Bernhard Pospichal

Remote sensing observations of the West African monsoon

Bernhard Pospichal

Remote sensing observations of the West African monsoon

Diurnal to annual variability of the lower atmospheric layers in the West African monsoon system: A comprehensive view by remote sensing observations

Südwestdeutscher Verlag für Hochschulschriften

Impressum / Imprint
Bibliografische Information der Deutschen Nationalbibliothek: Die Deutsche Nationalbibliothek verzeichnet diese Publikation in der Deutschen Nationalbibliografie; detaillierte bibliografische Daten sind im Internet über http://dnb.d-nb.de abrufbar.
Alle in diesem Buch genannten Marken und Produktnamen unterliegen warenzeichen-, marken- oder patentrechtlichem Schutz bzw. sind Warenzeichen oder eingetragene Warenzeichen der jeweiligen Inhaber. Die Wiedergabe von Marken, Produktnamen, Gebrauchsnamen, Handelsnamen, Warenbezeichnungen u.s.w. in diesem Werk berechtigt auch ohne besondere Kennzeichnung nicht zu der Annahme, dass solche Namen im Sinne der Warenzeichen- und Markenschutzgesetzgebung als frei zu betrachten wären und daher von jedermann benutzt werden dürften.

Bibliographic information published by the Deutsche Nationalbibliothek: The Deutsche Nationalbibliothek lists this publication in the Deutsche Nationalbibliografie; detailed bibliographic data are available in the Internet at http://dnb.d-nb.de.
Any brand names and product names mentioned in this book are subject to trademark, brand or patent protection and are trademarks or registered trademarks of their respective holders. The use of brand names, product names, common names, trade names, product descriptions etc. even without a particular marking in this work is in no way to be construed to mean that such names may be regarded as unrestricted in respect of trademark and brand protection legislation and could thus be used by anyone.

Verlag / Publisher:
Südwestdeutscher Verlag für Hochschulschriften
ist ein Imprint der / is a trademark of
OmniScriptum GmbH & Co. KG
Heinrich-Böcking-Str. 6-8, 66121 Saarbrücken, Deutschland / Germany
Email: info@svh-verlag.de

Herstellung: siehe letzte Seite /
Printed at: see last page
ISBN: 978-3-8381-1574-0

Zugl. / Approved by: Köln, Universität zu Köln, Diss., 2009

Copyright © 2010 OmniScriptum GmbH & Co. KG
Alle Rechte vorbehalten. / All rights reserved. Saarbrücken 2010

La science est le tronc d'un baobab
qu'une seule personne ne peut embrasser

Proverbe béninois

Abstract

Weather and climate in West Africa are determined by the pronounced contrast between tropical, moist air masses over the Gulf of Guinea in the south and the dry desert climate over the Sahara in the north. The sharp border between the two air masses exhibits a pronounced meridional annual cycle and follows the path of the sun northward. This circulation system is called "West African monsoon". In the past, the knowledge about the factors that control the monsoon and its strength was limited due to the small number of high-quality observations. Therefore, little is known about the reasons for the significant decline of annual rainfall over the Sahel area during the past 40 years which represents the most pronounced climatic signal worldwide. During the past few years, intensive atmospheric observations were performed in the framework of the international project "African Monsoon Multidisciplinary Analyses" (AMMA) in order to obtain high-quality data and to improve the process understanding.

This work gathered and analyzed ground-based remote sensing observations which were performed in Benin and Niger during the AMMA field campaigns. These data give an insight into diurnal and annual cycles of atmospheric parameters, such as water vapor, temperature profiles, cloud cover, cloud liquid water content, or wind with a temporal resolution never reached before. Particular attention is paid to the atmospheric water which is recognized to be a critical parameter for many other atmospheric variables, e.g. the vertical temperature distribution, the long-wave radiation balance, and many more.

The new type of observations revealed a diurnal cycle of the position of the Intertropical Discontinuity (ITD) prior to the start of the rainy season. The ITD represents the convergence zone at the surface between the dry and hot north-easterly trade winds and the moist and cooler south-westerly monsoon flow. Associated with this sharp front between the two air masses, strong water vapour changes occurred which could be observed in detail. This data set was then used for a comparison with the mesoscale atmospheric model Méso-NH which was run for a case study in April 2006. It is shown that the model reproduces the observed processes quite well, despite the low number of in-situ data which were assimilated in the model. Therefore, the model is suited to describe the processes in the lower atmospheric layers around the ITD.

Zusammenfassung

Wetter und Klima in Westafrika sind geprägt durch den ausgeprägten Kontrast zwischen tropisch-feuchten Luftmassen über dem Golf von Guinea im Süden und dem trockenen Wüstenklima der Sahara im Norden. Die meist scharfe Grenze zwischen den beiden

Luftmassen weist einen ausgeprägten meridionalen Jahreszyklus auf und folgt dem Lauf der Sonne nach Norden. Dieses Zirkulationssystem wird "Westafrikanischer Monsun" genannt. Das Wissen über die Faktoren, die den Monsun und seine Stärke steuern, war in der Vergangenheit nicht zuletzt durch die geringe Zahl der Beobachtungen beschränkt. Daher war über die Ursache für das deutlichste Klimasignal der letzten 40 Jahre weltweit, nämlich den signifikanten Rückgang der Jahresniederschläge über dem Sahel, bislang noch wenig bekannt. Im Rahmen des internationalen Projekts "Afrikanischer Monsun Multidisziplinäre Analysen" (AMMA) wurden in den vergangenen Jahren umfangreiche Beobachtungen durchgeführt, um diesen bislang bestehenden Mangel an qualitativ hochwertigen Daten zu beheben und dadurch das Prozessverständnis zu verbessern.

Diese Arbeit sammelt und analysiert vorwiegend bodengebundene Fernerkundungsmessungen, die während der AMMA Feldmesskampagnen in Benin und Niger durchgeführt wurden. Diese Daten geben einen Einblick in jährliche und tägliche Zyklen verschiedener atmosphärischer Parameter wie Wasserdampf, Temperatur, Wolkenbedeckung, Flüssigwassergehalt der Wolken und Wind mit einer zuvor nie erreichten zeitlichen Auflösung. Der atmosphärische Wassergehalt wird dabei detailliert untersucht, da dieser ein kritischer Parameter für viele andere atmosphärische Größen, wie etwa die vertikale Temperaturverteilung oder die langwellige Strahlungsbilanz ist.

Mit den verwendeten neuen Methoden konnte vor dem Beginn der Regenzeit ein Tageszyklus der Lage der Innertropischen Front (ITF) beobachtet werden. Die ITF stellt die Konvergenzzone zwischen den trocken-heißen Nordostpassaten und der feuchten und kühleren südwestlichen Monsunströmung dar. An dieser scharfen Luftmassengrenze traten starke tägliche Wasserdampfschwankungen auf, die mit den zur Verfügung stehenden Messgeräten detailliert beobachtet werden konnten. Für eine Fallstudie im April 2006 wurde dieser Datensatz mit den Ergebnissen des mesoskaligen Wettermodells Méso-NH verglichen. Dabei zeigte sich, dass das Modell trotz der geringen Anzahl der In-situ Beobachtungen, die in die Modellanalyse eingingen, die beobachteten Phänomene gut wiedergibt und daher zur Beschreibung der Prozesse um die ITF in den unteren Atmosphärenschichten geeignet ist.

Contents

1	**Introduction**	**7**
2	**West African climate**	**10**
3	**AMMA - African Monsoon Multidisciplinary Analyses**	**16**
	3.1 Implementation of the AMMA programme	17
	3.1.1 WP 4.2—Observation strategy during AMMA	17
	3.1.2 WP 2.1—Convection and atmospheric dynamics	20
	3.2 Measurement sites during AMMA	20
	3.2.1 Nangatchori/Djougou	21
	3.2.2 Niamey	24
4	**Instrumentation and measurement principles**	**26**
	4.1 Microwave radiometry	26
	4.1.1 Theory	26
	4.1.2 HATPRO system	28
	4.2 Lidar Ceilometer	35
	4.3 Micro Rain Radar	36
	4.4 Additional instrumentation	39
	4.5 Data availability	41
5	**Analysis of the annual and diurnal cycle of water cycle parameters**	**43**
	5.1 Annual cycle	43
	5.2 Diurnal cycle	56
6	**Detailed investigation of the ITD diurnal cycle over Nangatchori**	**68**
	6.1 Introduction	68
	6.2 Mesoscale simulation	69
	6.2.1 Model description	69
	6.2.2 Simulation description	69
	6.3 Synoptic situation	70
	6.4 Dynamics and thermodynamics of the ITD	73
	6.4.1 Surface observations	73
	6.4.2 Ground-based remote sensing observations	75
	6.4.3 Satellite observations	79
	6.5 A spatio-temporal view of the ITD diurnal cycle from model results	82

6.6 Statistical analysis of April 2006	88
7 Conclusions and Outlook	**91**
Bibliography	**95**
List of Figures	**102**
List of Tables	**106**
Acronyms	**107**

1 Introduction

The mechanisms that influence the strength of the West African Monsoon (WAM) are still not well understood. The moisture contrast between the tropical climate along the Gulf of Guinea and the dry desert climate in the north results in a region with strong differences in rainfall throughout the years in this area. The amount of rainfall is a very critical and variable parameter for this climate region. During the past 40 years, a large decline (by up to 25 %) in precipitation has been observed, being the most pronounced climate signal worldwide in the last decades (Le Barbé et al., 2002). Periods of drought caused crop failures and widespread famines in the Sahel region.

The climate in West Africa is mainly characterized by the moist south-westerly monsoon flow in the southern part and the dry north-easterly trade winds (also called Harmattan) in the northern part. The interface between these two flows, called the Inter-tropical discontinuity (ITD), shows a distinct annual cycle (from 6 °N in January to 20 °N in July). Apart from the large-scale atmospheric and oceanic circulations, atmospheric humidity plays a key role in those processes that determine the strength of the monsoon. A significant part of the atmospheric water—whether in liquid or vapor phase—is located in the planetary boundary layer (PBL) and is furthermore an important parameter in surface exchange processes. For this reason, detailed observations of atmospheric conditions in the lowest part of the atmosphere are essential to obtain a comprehensive view of the monsoon system. Furthermore, ground-based observations can help to validate atmospheric models and possibly lead to the development of improved parameterization schemes.

In order to broaden the knowledge of the processes controlling the West African Monsoon, the international project "African Monsoon Multidisciplinary Analyses" (AMMA) (Redelsperger et al., 2006) was launched, combining a wide variety of ground-based, maritime, airborne, and satellite measurements. One of the main tasks was to establish a ground-based remote sensing observation network over this area. To reach this goal, three so-called supersites were established in different climate zones (Lebel et al., 2009) where a comprehensive view of atmospheric, aerosol, hydrological, and surface-exchange observations should be provided by a suitable set of instrumentation.

Observations are rather sparse in West Africa. Until recently, for many countries only standard surface meteorology data were available. Especially the network of upper-air observations was very poor: only three stations, i.e. Dakar (Senegal), Bamako (Mali), and Niamey (Niger), provided regular radiosonde data over an area of about 5 Mill. km^2 (Parker et al., 2008). The radiosounding network has been extended as part of AMMA which is vital for an improvement of atmospheric modeling and weather forecasts. During the intense observing campaigns in 2006 twelve radiosonde stations operated regularly in this area, but even then, most of the stations performed soundings only two to four times a day at the main synoptic hours. In the past, some campaigns, e.g. HAPEX-Sahel in 1992 (Dolman et al., 1997), and Jet 2000 (Thorncroft et al., 2003) have provided more detailed observations by aircraft, enhanced radiosoundings, pilot balloons as well as in situ surface stations. A radiosonde campaign with twice daily ascents was performed at Parakou, Benin from April to October 2002 in the frame of the Impetus project. These data have been used by Fink et al. (2006) to study atmospheric profiles before and after different types of rainfall events, and Schrage et al. (2007) investigated nocturnal stratiform cloudiness during monsoon time. Since all these campaigns were confined to limited time intervals, Parker et al. (2005) note the complete lack of longer term measurements with high temporal resolution in that region. This deficit can also not be closed through satellite observations because these measurements do not resolve the atmospheric boundary layer adequately.

This thesis focuses on observations at the supersite Nangatchori, because the operation of several instruments there was part of this work. The main emphasis was on continuous thermodynamic monitoring of the lower troposphere in 2006. This was performed with high temporal resolution by a novel ground-based microwave radiometer, the Humidity And Temperature PROfiler HATPRO (Rose et al., 2005). HATPRO is able to observe temperature profiles with high vertical resolution in the atmospheric boundary layer (Crewell and Löhnert, 2007) in addition to the standard products IWV (integrated water vapor), LWP (cloud liquid water path), and full troposphere temperature and humidity profiles. To our knowledge, this has been the first time that such a microwave radiometer was used in West Africa for monitoring the lower troposphere. These data are completed by additional instruments at Nangatchori which include a lidar ceilometer, vertical pointing Doppler rain radar, measurements of temperature, humidity, and wind on a tower in 5 levels up to 6 m, detailed in situ aerosol observations, wind profiler, and ozone lidar. A similar set of instruments was installed in Niamey (Niger), being 400 km north of Nangatchori and hence in a much drier climate zone. Both at Niamey and Djougou, most of the instruments were deployed over the entire year 2006 allowing the examination of the full annual cycle.

This work gives a comprehensive analysis of atmospheric variables, such as water vapor, temperature, cloud cover, or LWP. It is shown that the highly variable atmospheric water vapor content is one of the main drivers of the atmospheric variability and has a large influence on many other parameters, such as temperature profiles or the longwave radiation balance. Through a combination of different remote sensing measurement methods, it is possible to describe the PBL development for different seasons in detail. Thanks to the high temporal resolution compared to radiosondes, it is possible to present the diurnal cycle of the ITD at the transition to the monsoon season for the first time.

1 Introduction

This phenomenon was captured by a microwave profiler, a wind profiler, a lidar ceilometer and many other surface observations which enabled to gain a comprehensive view of the processes around the ITD and a statistical examination on PBL properties.

The first part of this thesis consists of a short overview of the relevant climatic features in connection with the West African monsoon and the low-level circulation is given (chapter 2). In a further step, the project AMMA and the supersites will be presented (chapter 3). The instrumentation is described in chapter 4 with a special focus on the ground-based remote sensing instruments which were deployed in the frame of this work at the Nangatchori supersite.

Chapter 5 is dedicated to the detailed examination of the ground-based remote sensing data regarding both annual and diurnal variabilites of the lower atmosphere's conditions. For example, the relationship between the long-wave radiation balance and IWV as well as the diurnal temperature range will be presented. The diurnal distribution of cloud cover and its different evolution throughout the seasons is another point which will be addressed.

In the next step (chapter 6), a case study around the onset of the monsoon is presented when the ITD was located over the Nangatchori area. A diurnal cycle of the ITD position was observed in combination with a large water vapor contrast and a nighttime northward transport of moist air in low levels. For this case study, the mesoscale atmospheric model Méso-NH has been run with 10 km horizontal resolution. A comparison of observed profiles and model results will be shown and the model performance is evaluated.

The thesis will conclude with a summary of the main results of this work, especially regarding the detailed description of the annual and diurnal cycle of PBL parameters over Nangatchori and Niamey. A short outlook will suggest longer term observations to capture inter-annual fluctuations in the atmosphere and will discuss if the use of additional ground-based remote sensing instruments would provide even more insight into the PBL processes in the monsoon system.

2 West African climate

West African [1] climate is characterized by a sharp contrast between the almost constantly humid conditions at the Guinea Coast and the arid conditions in the Sahara to the north. This difference is caused by the global circulation system with the upward branch of the Hadley circulation in the equatorial region and the subsiding air masses in the subtropics further to the north. The region with the upward motion is called the Intertropical Convergence Zone (ITCZ).

The climatic conditions of the African continent according to the Köppen-Geiger classification (Köppen and Geiger, 1928) are presented in Fig. 2.1. The most important climate zones for West Africa are explained briefly. Some regions close to the Guinea Coast are in the zone of **equatorial monsoon climate (Am)** with rainfall in all months of the year, but showing a distinct rainfall minimum in the winter season. Large parts between 6 °N and 12 °N are within the **equatorial savanna climate with dry winter (Aw)** with two pronounced seasons (dry and wet). A small band of **hot steppe climate (BSh)** follows north of the Aw zone in the Sahel area between about 12 °N and 15 °N. North of that, the Sahara with its **hot desert climate (BWh)** extends over more than 15 degrees of latitude. Details on the Köppen-Geiger climate classification can be found in Kottek et al. (2006).

The sharp moisture gradient between the humid tropical climate along the Gulf of Guinea in the south and the dry desert climate in the north causes a large rainfall variability throughout the years in the Sahel area. A slight meridional monsoon shift can considerably change the amount of rainfall which is a very critical parameter for this climate region. During the past 40 years, a large decline in precipitation has been observed in the Sahel area, which is the most pronounced climate signal worldwide in the last decades (Le Barbé et al., 2002). For example, the annual mean precipitation for Niamey is 654 mm between 1950 and 1967, and only 495 mm for the period 1968-1989 (Lebel et al., 1997). This is a decline by nearly 25 %. Similar observations are made over the whole Sahel region. After 1990, the rainfall amounts have slightly recovered, but are far from reaching the values from the 1950s and 60s (Lebel and Ali, 2009). Further to the south, rainfall is much more abundant, but in Nangatchori, too, a decline from 1340 mm (1950–70) to 1208 mm (1970–90) was observed.

---

[1] West Africa geopolitically encompasses those countries that lie between the Atlantic Ocean in the south and west, the Sahara in the north, and an imaginary north-south boundary line at approximately 10 °E

2 West African climate

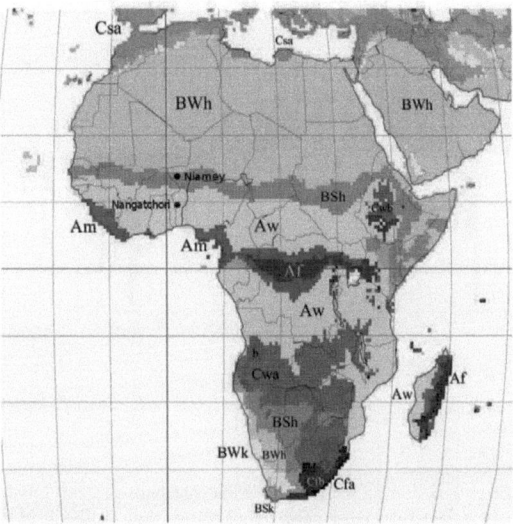

Figure 2.1: Climate zones of Africa (Köppen-Geiger classification) after Kottek et al. (2006).

To visualize this dramatic decline of rainfall, Ali and Lebel (2009) calculated a standardized precipitation index (SPI) for the Sahel region (10 °N–17.5 °N, 22.5 °W–17.5 °E). The SPI is defined as the arithmetic mean of the normalized annual precipitation recorded at several stations over the region. The standard deviation computed at each station over a period of reference is used as the normalizing factor. Fig. 2.2 shows that after 1967 only three years were wetter than the average of 1905–2006.

The dominant circulation feature in West Africa is the annual cycle of the ITCZ over the continent. During the whole year, the convergence zone at the ground lies further north than the ITCZ above. This low-level convergence zone is called Intertropical Discontinuity (ITD) or monsoon trough and separates the moist and relatively cool south-westerly monsoon flow from the dry and warm north-easterly Harmattan flow (Hastenrath, 1985; Bou Karam et al., 2008; Flamant et al., 2009). Fig. 2.3 presents the mean position of the ITCZ at 15 °N, and the ITD at 21 °N as well as the related weather systems in August.

As the position of the ITCZ follows the path of the sun northward in March/April, moist air masses start to propagate from the Gulf of Guinea to inland. This defines the begin of the first rainy season over the Guinea Coast region south of 10 °N (Sultan and

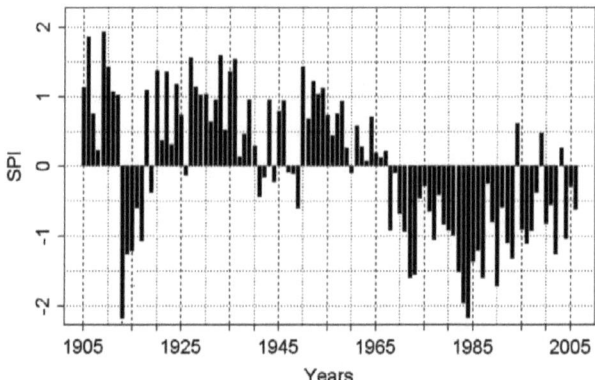

Figure 2.2: Standardized precipitation index for the Sahel area. Source: Ali and Lebel (2009).

Janicot, 2003). During the following months, the ITD continues to move inland (Lothon et al., 2008), reaching its northernmost climatological position at about 21 °N in July and August (Sultan et al., 2007). The southward retreat starts in September and takes place much quicker than the northward move. In November, the ITD reaches already its southernmost position close to the Guinea Coast (4 − 6 °N).

The shape of the monsoon trough (or ITD) in Fig. 2.3 is determined by differences in density between the dry and hot continental air from the desert and the tropical maritime air from the Atlantic which is slightly cooler than the hot air. At low levels, the contrast between the two air masses is often very sharp. At the ground, the ITD position can be determined by the 15 °C dewpoint isoline (Buckle, 1996). The ITD position does not move gradually over several months and rarely remains stationary for a longer period, but there are many processes that impact the ITD at various time scales: from nighttime low-level jets (Flamant et al., 2009) to multi-day pulsations with cycles of about 5 days (Couvreux et al., 2009). These pulsations are associated with increased meridional low-level winds and bring additional moisture to the north. Bock et al. (2008) describe the seasonal evolution of the diurnal water vapor cycle for several locations in West Africa using Global Positioning System (GPS) observations. They show that for each of these locations, strong diurnal cycles occur only during one or two distinct months in connection with the ITD. The mean ITD position along the 0 ° meridian for the months April–October is shown in Fig. 2.4.

The diurnal cycle of the ITD is one of the atmospheric processes which play a crucial role in the WAM system. With the associated (mainly nocturnal) low-level winds, the

2 West African climate

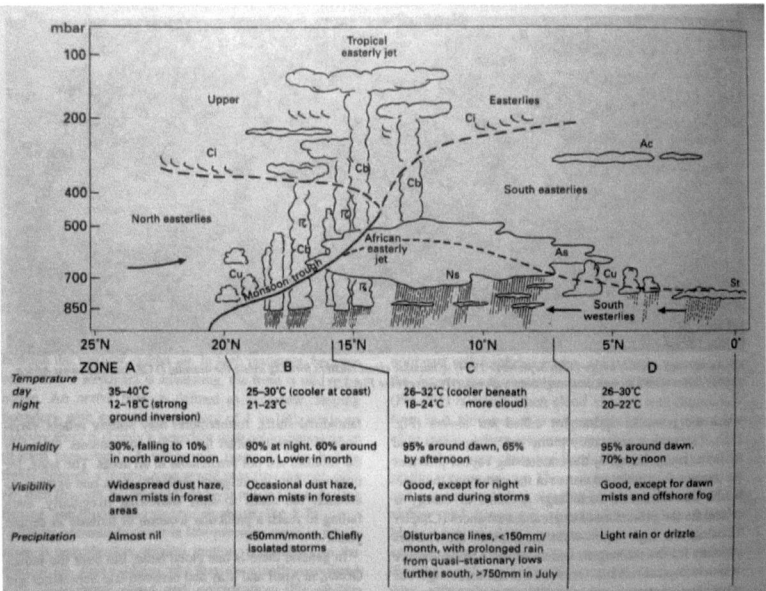

Figure 2.3: Meridional section across West Africa in August showing the position of the main airstreams, the monsoon trough, and the principal weather zones with the ITCZ at 15 °N and the ITD at 21 °N. Source: Buckle (1996).

diurnal variation of the ITD position has been recognized in several previous studies as a key factor for the northward transport of moisture (e.g. Sultan et al. (2007); Lothon et al. (2008). Parker et al. (2005) give already a detailed overview of previous research. Therefore, in this work only a brief summary is presented in the following. During daytime the heat low over the Sahara intensifies with a pressure minimum in the afternoon. As the convective boundary layer grows during the day, vertical mixing prevails and the horizontal flow is rather weak. In the late afternoon when solar heating ceases turbulence stops rapidly, and the lower atmosphere is able to respond to the pressure gradient force induced by the heat low. The low level southerly flow intensifies over night and its edge continuously moves northward. This nocturnal meridional flow is responsible for the advection of moist air in low levels further inland and forms the main moisture source for summertime convection in the Sahel. In higher regions (around 700 hPa) there is a dry return flow. During daytime, low level humidity drops as dry air from above is mixed down in the developing convective boundary layer. In Chapter 6, a case study at the onset of the monsoon season analyses the diurnal cycle of the ITD over Benin before the

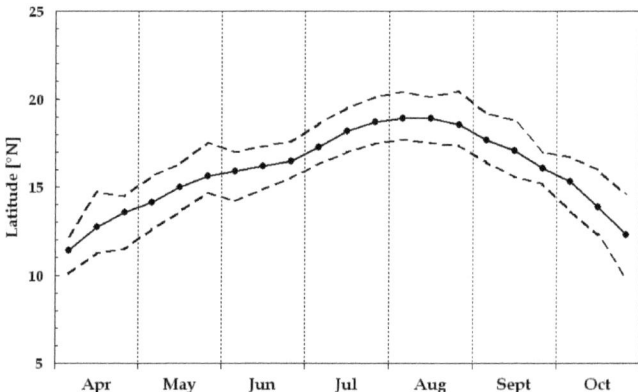

Figure 2.4: Solid line: 10-day average ITD position along the 0° meridian (mean over the years 1988–2005). Dashed lines indicate the maximum and minimum position during this period. Source: NOAA Climate Prediction Center, African ITCZ Monitoring Project (http://www.cpc.noaa.gov/products/fews/ITCZ/itcz.shtml).

monsoon onset in April and gives an overview over the related atmospheric processes.

Similar observations of a diurnal cycle between low-level moist and dry air masses associated with heat lows were made in Northern Australia. Racz and Smith (1999) performed numerical investigations where they show deep convective mixing during daytime, the development of a nighttime low-level jet and nocturnal frontogenesis in the transition zone of the different air masses. Furthermore, they speculate that these features are likely to occur elsewhere under similar climatic conditions.

Nearly all year round, nocturnal low-level jets occur with more or less intensity over the region and peak up to $15 - 20\,\mathrm{ms}^{-1}$ below 500 m above ground level (AGL) (Lothon et al., 2008). The origin of these low-level jets is already widely known. During daytime, convective turbulent mixing is large and limits horizontal transport of air masses. In the late afternoon, the heat flux decreases, and the circulation responds to the surface pressure gradient force imposed by the heat low without being slowed down by friction. The winds are accelerated towards the low pressure during the night, reaching their maximum speed in the morning hours. Due to the proximity to the equator, the influence of Coriolis force is rather small. The speed and depth of the low-level jet depend on a range of factors, including baroclinity, terrain slope, surface cooling rates, and many more (e.g. Blackadar, 1957, or Garratt, 1994). However, over West Africa the strength

2 West African climate

of the nocturnal jet as well as the sharpness of water vapor and temperature gradients across the ITD during its northward displacement have been only little observed until now. An exception are several days of observations in Mali and Niger during the early monsoon season in June 2006 (Flamant et al., 2009) where sharp low level water vapor and wind gradients associated with the ITD were observed by dropsondes. At that time of the year, the ITD had almost reached its northernmost position.

3 AMMA - African Monsoon Multidisciplinary Analyses

The AMMA project has been launched in 2004 as an international project to investigate the West African climate in an extent which had never been done before. It is funded by both the European Union and national funding agencies mainly in France, the United Kingdom and the US (Redelsperger et al., 2006).

The dramatic change from the abundant rainfall in the 1950s and 60s to much drier conditions from the 70s to the 90s over entire West Africa was the strongest trend in rainfall on the planet of the 20th century (Le Barbé et al., 2002). Marked inter-annual variations in recent decades have resulted in extremely dry years with devastating environmental and socio-economic impacts. With a large rural population depending on rain fed agriculture, the abrupt decrease of water resources has been devastating to both populations and economies. It is AMMA's aim to provide the African decision makers with improved assessments of similar rainfall changes which are likely to occur during the 21st century due to natural fluctuations and as a result of anticipated global climate change. An essential step in that direction is to improve our ability to forecast the weather and climate in the West African region.

The international AMMA programme has three overarching goals[1]:

- To improve our understanding of the WAM and its influence on the physical, chemical and biological environment regionally and globally.

- To provide the underpinning science that relates climate variability to issues of health, water resources, food security, and demography for West African nations and defining relevant monitoring and prediction strategies.

- To ensure that the multidisciplinary research is effectively integrated with prediction and decision making activity.

---

[1]Source: AMMA project website (http://www.amma-international.org)

3 AMMA - African Monsoon Multidisciplinary Analyses

3.1 Implementation of the AMMA programme

The large number of more than 40 participating organisations from Europe, Africa and North America made it essential to structure the project adequately. Therefore, the implementation of AMMA was split into four main areas, namely research, demonstration, education and management activities.

This work is focused on the atmospheric research activities within AMMA. In order to reach the goals of an improved understanding of the WAM system, this wide part of research activities was structured into different areas:

1. Integrative science: Studies of large-scale atmospheric processes that influence the WAM as well as fundamental research on the water cycle and surface-atmosphere feedbacks over West Africa.

2. Process studies: Small to medium-scale studies of atmospheric dynamics, convection, aerosols, chemical processes as well as oceanic studies.

3. Impact studies: Influence of changes in the WAM system on land productivity, water resources, and health impacts, as well as adaptation of humans to changed conditions.

4. Tools and methods: Collection of various data sets for the studies in items 1–3, such as organizing the field campaigns, establishing a database, preparing satellite data, or performing model evaluation and data assimilation.

These four areas were split into different work packages (WP). The activities of the author have mainly been focused on WP 4.2 (Field campaigns, as a part of "Tools and methods") and WP 2.1 (Convection and atmospheric dynamics, part of "Process studies"). These two points of the AMMA project will be presented in more detail.

3.1.1 WP 4.2—Observation strategy during AMMA

The implementation of observations in West Africa was a challenging task for the WP 4.2, as such a huge amount of instruments had never been operated in this region before and already basic logistic issues, such as electricity or transportation were often problematic. To organize, coordinate, and document the complex field programs, 10 task teams were established according to the different observation strategies like the type of observation (ocean, atmosphere, hydrology), the platform (land, air, sea, satellite), and the timeframe of the measurements.

3.1 Implementation of the AMMA programme

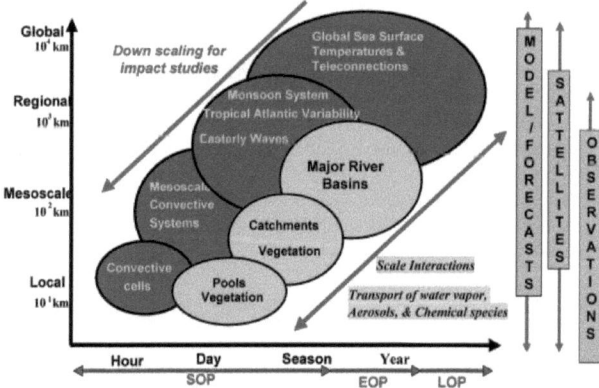

Figure 3.1: Space and time scales in AMMA. Source: Redelsperger et al. (2006).

The AMMA field program aimed to provide data to document all the components of the WAM system on four interacting temporal and spatial scales (see Fig. 3.1).

The temporal scale of observations was divided into three different periods to cover all relevant topics:

The Long-term Observing Period (LOP) was concerned with observations of two types: the historical observations to study inter-annual to decadal variability and additional long term observations (2001–2010) to document and analyse the interannual variability of the WAM. Given the great variability of climatic conditions from year to year in this region, observations are needed over several annual cycles in order to gain a proper view on the diversity of the seasonal cycles around a mean state. LOP observations are still conducted from three main mesosites, with enhanced observations of rainfall, hydrology, surface and subsurface conditions. These stations, one being Nangatchori, will be described in more detail in section 3.2.

The Enhanced Observing Period (EOP) covered a three-year period (2005–2007). Its main objective was to document the annual cycle of the surface conditions and atmosphere over a meridional transect. The strategy of the EOP was to enhance the observations such as extra radiosoundings, new surface flux measurements and ground-based remote sensing observations (radars, wind profilers, etc.). In addition, increased vegetation and oceanic observations were taken.

The Special Observing Periods (SOP) focused on detailed observations of specific

3 AMMA - African Monsoon Multidisciplinary Analyses

processes and weather systems during the dry season (SOP 0) and at various key stages of the rainy season during three periods in 2006 (SOP 1, 2, and 3).

- **SOP 0—Dry season** (January to February 2006): The main objective of SOP 0 was to characterize aerosol properties at the local scale resulting from mixing and transport and their impact at the regional scale. This was performed by observations in the Benin/Niger area (mineral dust and biomass burning aerosols) and in the Dakar area (transport of Saharan dust onto the Atlantic ocean) by both ground-based and aircraft observations.

- **SOP 1—Monsoon onset** (15 May to 30 June 2006): Its objective was to provide the necessary observations required to support analysis and understanding of the seasonal evolution of the atmospheric low levels, the oceanic surface layer as well as their relationship with regional circulations and atmospheric and continental water budgets. Ground-based and aircraft measurements were centered in Benin, Burkina Faso, and Niger.

- **SOP 2—Peak monsoon** (1 July to 14 August 2006): This period was key to study the multiple scale interactions between the surface conditions, synoptic environment and the convective system. The role of mesoscale precipitating systems and trace gas transports were investigated. The instrument setup was similar to SOP 1 with additional aircrafts based in the Niger area.

- **SOP 3—Late monsoon** (15 August to 15 September 2006): The seasonally evolving surface conditions, low-level thermodynamic contrast and associated regional circulations continued to be observed during this SOP as the monsoon started to retreat southward, although greater emphasis in SOP 3 was given to the eastern tropical Atlantic jet downstream of West Africa. Therefore, the main focus of ground-based and aircraft measurements was at Dakar (Senegal).

In the frame of this thesis, three ground-based remote sensing instruments were operated in Nangatchori (Benin) over a full year's cycle (January 2006–January 2007) as part of the SOP and EOP measurements. First of all, the work included installation, maintenance, and dismantling of the instruments in Benin as well as the organization of the transport of instruments. In a second step, transfer and storage of data were organized. As there was no internet connection at the measurement site, a master student from Benin regularly saved the data. Furthermore, a detailed data quality control was performed (Pospichal and Crewell, 2007a,c).

3.1.2 WP 2.1—Convection and atmospheric dynamics

This WP is devoted to the study of the processes in the atmosphere which control the WAM and its downstream environment. The study of convection and its initiation due to different radiative forcing as well as distinct aerosol and humidity patterns play a key role in the better understanding the mechanisms of the WAM system. Especially the high variability in the lower atmospheric levels at the inhomogeneous lower atmospheric boundary and the related surface fluxes should be studied in detail. After the 2006 field campaigns, special "golden days" (case studies of interesting atmospheric phenomena for which a high number of observations are available) were selected.

Four sub-workpackages (sWP) have been defined in order to structure this widespread field. **sWP 2.1.A (Low levels)** deals with the evolution of the PBL, its diurnal, annual, and meridional variation, as well as the role of the PBL in convection initiation. **sWP 2.1.B (Heat Low)** mainly deals with the properties and the extent of the Saharan heat low, its influence on moist convection, and on the role played by aerosols, surface properties and topography. Furthermore, the interaction with African Easterly Waves and mid-latitudinal weather systems has been studied. The first goal of **sWP 2.1.C (Moist convection)** is to document precipitation, dynamics, microphysics, and lightning within convective systems observed during the SOP and its environment using the wide variety of ground-based and satellite observations. This shall lead to a better understanding and quantification of key processes involved by convection in particular concerning life cycle (initiation, dissipation), organization, precipitation efficiency (evaporation), microphysics, lightning activity, impact at large scale, coupling with the surface, and many more. **sWP 2.1.D (Large-scale dynamics)** aims to make a link between small scale atmospheric processes such as convection and the intra-seasonal variability of the WAM (> 10 days) which is treated by WP 1.1 (West African monsoon and the global climate).

The work of this thesis is mainly focused on the boundary layer processes (sWP 2.1.A) where detailed studies on the statistical description of the boundary layer are performed. Especially the annual variation of various atmospheric parameters in the course of the dry and wet seasons, as well as different diurnal cycles throughout the seasons have been studied in detail (see chapter 5 and Pospichal and Crewell (2007b)).

3.2 Measurement sites during AMMA

Before the AMMA campaign started, only standard surface observations were performed on a regular basis at most places in West Africa. Radiosondes were launched regularly only at a few stations (e.g., Dakar, Bamako, or Niamey), and no reliable upper air data were available for the Guinea Coast region. Therefore, one goal of AMMA was

3 AMMA - African Monsoon Multidisciplinary Analyses

to establish a network of radiosonde stations that performed regular soundings (Parker et al., 2008).

For a more detailed description of the water and energy cycle in the course of the year, a variety of in-situ and remote sensing instruments were gathered at three so-called supersites. These sites were located around the 0 °E meridian and covered three climate zones causing different surface and vegetation conditions. The southernmost and wettest site was at the Ouémé catchment around Djougou and Nangatchori, Benin (9.6 °N, 1.7 °E). Further north were the sites at Niamey, Niger (13.5 °N, 2.1 °E) and Gourma, Mali (16 °N, 1.5°W). The monsoon flow usually arrives in Djougou in April, whereas Gourma does not become fully influenced by the moist air masses before June (Bock et al., 2008). At all three supersites, surface meteorology and hydrology measurements were performed, whereas extended ground-based remote sensing observations were made only at the two stations in Benin and Niger. In the following, these two stations and the instrumentation deployed there will be presented in more detail since the results presented in this work mainly originate from these stations.

3.2.1 Nangatchori/Djougou

The site of Nangatchori (Fig. 3.2) is located 10 km south-east of Djougou at 415 m above mean sea level. Djougou is the capital city of a province with the same name in central Benin, about 300 km north of the Atlantic Ocean coast (Gulf of Guinea). This area lies in a tropical climate with a dry season between October and March and a wet season between April and September (also known as Savanna climate or tropical wet and dry climate, see chapter 2). Temperatures rarely drop below 20 °C; the extreme temperature range is 15 to 42 °C. The measurement site is surrounded by manioc fields and some shrubbery in the east of the small village of Nangatchori. The site was established in early 2005 to serve as an AMMA measurement site. Although the site was permanently guarded for security reasons, the infrastructure to operate the various instruments without skilled operators was quite problematic due to the unreliable power supply and missing phone and internet connection. In terms of electricity, the whole town of Djougou was only supplied by one diesel generator which frequently broke down. Therefore much effort had to be put into preparing the instruments for the hot and partially humid climate as well as for the unreliable power supply.

The instrumentation at this site combined a variety of in-situ and ground-based atmospheric remote sensing instruments with hydrological measurements from many different project partners in AMMA. The common goal was to characterize the atmospheric state as well as the surface exchange processes as detailed as possible. Many instruments were deployed for more than one year in order to get a cycle of an entire year. Some special measurements were confined to SOP periods.

3.2 Measurement sites during AMMA

Figure 3.2: Nangatchori site in January 2006. Left: Tower with turbulence and flux measurements. Center: Lidar Ceilometer. Right: HATPRO microwave radiometer

In order to obtain a characterization of the boundary layer over the area, a microwave profiler (HATPRO), a lidar ceilometer and a Micro Rain Radar (MRR) were installed in the frame of this thesis. They operated in an autonomous mode over a full year from January 2006 to January 2007 (Pospichal and Crewell, 2007a). Table 3.1 gives an overview of some characteristics for these instruments.

For a complete characterization of the atmosphere and the surface exchange processes many other instruments were located at this site. Measurements of chemical compounds of the atmosphere as well as aerosol particle observations were performed. Apart from the lidar ceilometer, other instruments for optical remote sensing, one ozone and one aerosol lidar, as well as a sun photometer, were operating in Nangatchori. Aerosol studies were mainly performed during the dry season, focusing on both biomass burning and dust outbreaks. Biomass burning is a common agricultural practice in the dry season when remaining plants on the fields are burned down. Observations of biomass burning aerosols using lidar and in-situ measurements are presented in Pelon et al. (2008) and Mallet et al. (2008). Due to the frequent moist air outbreaks from the south early in the year, larger dust events were rare in 2006. However, the only major event around 9 March brought a considerable dust load and was described in detail by Tulet et al. (2008). During the monsoon season, the biogenic volatile organic compound composition has been studied at the site in combination with a detailed description of the plants around the measurement site (Saxton et al., 2007).

3 AMMA - African Monsoon Multidisciplinary Analyses

Table 3.1: Instruments in Nangatchori in 2006, used in this study.

Instruments	Spectral range	Measured parameters	Resolution, Accuracies
Microwave radiometer RPG-HATPRO	7 channels at H_2O absorption line (22.24–31.4 GHz) 7 channels at O_2 absorption complex (51.26–58.0 GHz)	14 channel microwave brightness temperatures (zenith obs., every 15 min. elevation scanning) Derived parameters: • IWV (total column atmospheric water vapor content) • LWP (atmospheric liquid water path) • Temperature (T) and humidity (q) profiles	Temporal resolution: 2 sec. for zenith observations, 15 min. for elevation scans Accuracies: • IWV: $< 1\,\mathrm{kg\,m^{-2}}$ • LWP: $20\text{--}30\,\mathrm{g\,m^{-2}}$ • T-Profiles: 0.5 K at 500 m AGL, decreasing to 2 K at 5 km AGL (Crewell and Löhnert, 2007) • q-Profiles: $0.6\text{--}1.2\,\mathrm{g\,m^{-3}}$, max. 2 independent layers (Löhnert et al., 2009)
Lidar Ceilometer Vaisala CT25K	$\lambda = 905$ nm	Vertical backscatter profiles up to 7 km AGL Cloud base height (up to three layers can be detected)	Temporal resolution: 15 sec. Vertical resolution: 30 m
Micro Rain Radar Metek	24.1 GHz	Vertical Doppler spectra Derived parameters: • Drop size distribution • Fall velocity • Rain rate	Temporal resolution: 10 sec. Vertical resolution: 160–200 m[a], vertical range: 4800–6000 m (depending on measurement mode).
UHF Doppler wind profiler Degreane PCL-1300	1.29 GHz	Vertical profiles of wind speed and direction up to 3–5 km AGL	Temporal resolution: 5 min. Vertical resolution: 75 m
GPS receiver in Djougou (10 km west of Nangatchori)		Zenith tropospheric wet delay Derived parameter: • Integrated water vapor	Temporal resolution: 15 min. Accuracy $0.5\,\mathrm{kg\,m^{-2}}$ (Bock et al., 2008)

[a]The vertical resolution of MRR was changed in June 2006 from 200 to 160 m.

3.2 Measurement sites during AMMA

Data of wind profiles were gathered by Very High Frequency (VHF) and Ultra High Frequency (UHF) radars from April 2006 to September 2007. These instruments were operated by Centre National de Recherches Météorologiques (CNRM). Furthermore, disdrometers and some pluviometers have been installed for hydrological and precipitation studies, supporting the X-Port radar at Djougou. A study of drop size distributions observed in Nangatchori is presented by Moumouni et al. (2008) using optical disdrometer data. They find a bimodal drop size distribution for squall lines and mesoscale convective systems with convective and stratiform regions.

A six meter tower for eddy correlation measurements (water vapor and CO_2 fluxes) was constructed. It was equipped with temperature and humidity sensors as well as anemometers in different levels. The temperature sensors which are used for the case study in chapter 6 of the present work were are located 1.2 m, 2.5 m and 4 m AGL. Data are available with a 15 min temporal resolution. Wind direction and wind speed measurements were performed at four levels. Wind and temperature data from the tower exhibit large data gaps during the investigation period of this study. Therefore observations from an automatic weather station which was operated by the Institut de Recherche pour le Développement (IRD) in Djougou (10 km north-west of the measurement site) with a temporal resolution of 15 min are used in addition.

3.2.2 Niamey

At the airport of Niamey, the capital city of Niger (13.5 °N, 2.1 °E, approximately 400 km north of Djougou), another AMMA supersite was set-up (Fig. 3.3). The instrumentation at this location consisted mainly of the mobile facility of the Atmospheric Radiation Measurement (ARM) program of the US Department of Energy (Miller and Slingo, 2007) which was deployed there between December 2005 and January 2007. The measurement setup consisted of a variety of active and passive remote sensing instruments (lidars, wind profiler, microwave radiometers, cloud radar, radiation measurements) as well as surface measurements (aerosol characteristics, eddy covariance station, standard weather station). In addition, a C-Band Doppler weather radar of the Massachusetts Institute of Technology (MIT) operated in Niamey during the wet season of 2006.

3 AMMA - African Monsoon Multidisciplinary Analyses

Figure 3.3: ARM mobile facility (AMF) in Niamey in November 2005 (shortly after deployment). In the front is the aerosol stack and the containers, in the background the AMF instrument field. Courtesy: U.S. Department of Energy's Atmospheric Radiation Measurement Program.

4 Instrumentation and measurement principles

During the AMMA campaign, a variety of ground-based remote sensing observations were performed in West Africa, mainly at the supersites in Niamey and Nangatchori. Since this work is based on many of these observations, the essential information on the main instruments used in this study and their measurement principles will be given in the following section.

4.1 Microwave radiometry

The use of passive microwave radiometry for atmospheric observations became more and more popular during the recent years. Several specific emission lines of atmospheric constitutents (e.g., water vapor, oxygen) and the continuum emission of cloud liquid water in the centimeter and millimeter regions of the electromagnetic spectrum allow to obtain information on water vapor, cloud water and temperature in the atmosphere, using both ground-based and spaceborne receivers. The semi-transparency of clouds in this frequency range makes it possible to observe under nearly all atmospheric conditions.

This work presents the results of ground-based microwave observations for which different instruments were used. In Nangatchori, a 14-channel radiometer (HATPRO, Rose et al. (2005)), manufactured by Radiometer Physics GmbH, was installed. At the AMF site in Niamey, two Radiometrics microwave radiometers were deployed, one 12-channel profiler and one two channel radiometer for IWV and LWP observations (Ware et al., 2003; Turner, 2007). The exact frequencies of these instruments are listed in Tab. 4.1.

4.1.1 Theory

In a non-scattering medium (which is valid under normal atmospheric conditions for frequencies below 100 GHz, Janssen (1993)), the radiation intensity is given by the Planck function B_ν which directly relates the physical temperature T to a specific intensity I_ν at a given frequency ν for a blackbody emitter with an emissivity $\epsilon = 1$:

4 Instrumentation and measurement principles

Table 4.1: Frequencies (in GHz) of the microwave radiometers deployed at Nangatchori and Niamey.

Instrument	HATPRO Nangatchori	12-channel Niamey	2-channel Niamey
K/K$_a$ band frequencies	22.24	22.035	23.8
	23.04	22.235	31.4
	23.84	23.835	
	25.44	26.235	
	26.24	30.0	
	27.84		
	31.4		
V band frequencies	51.26	51.25	
	52.28	52.28	
	53.86	53.85	
	54.94	54.94	
	56.66	56.66	
	57.3	57.29	
	58.0	58.8	

$$I_\nu = B_\nu(T) = \frac{2h\nu^3}{c^2(e^{h\nu/k_bT)}-1)} \tag{4.1}$$

with c being the speed of light, h Planck's constant and k_b Boltzmann's constant. For the microwave frequencies used here and typical temperatures in the atmosphere, the Rayleigh-Jeans approximation ($h\nu/k_bT \ll 1$) is valid. Therefore Eq. 4.1 can be simplified to

$$B_\nu(T) \approx \frac{2k_bT h\nu^2}{c^2} \tag{4.2}$$

which implies a direct proportonality between blackbody radiance and temperature. Thus, the radiance I_ν can be expressed as blackbody equivalent temperature or brightness temperature T_B (Eq. 4.3; i.e. the temperature T of a blackbody with the radiance I_ν.

$$T_B(\nu) = \frac{\lambda^2}{2k_b}I_\nu \tag{4.3}$$

With this definition, the solution of the radiative transfer equation for a ground-based microwave receiver in a non-scattering plane-parallel atmosphere can be written as (Janssen, 1993):

4.1 Microwave radiometry

$$T_B(\nu) = T_{B_{cos}}e^{-\tau} + \int_0^\infty \alpha(s)T(s)e^{-\tau(s)}ds \qquad (4.4)$$

Two sources contribute to this brightness temperature (BT): the cosmic background radiation $T_{B_{cos}} = 2.73K$ and the emission of atmospheric gases and hydrometeors at temperature $T(s)$ attenuated by absorption throughout the atmosphere, determined by the absorption coefficient $\alpha(s)$ and the optical depth

$$\tau(s) = \int_0^s \alpha(s')ds'. \qquad (4.5)$$

The typical absorption spectrum for the frequencies below 100 GHz is presented in Fig. 4.1. Transitions from different rotational states of the gas molecules cause the peaks in this distribution which lead to two main well defined absorption lines in this frequency range: the 22.235 GHz water vapor line and the oxygen absorption complex around 60 GHz consisting of more than 30 absorption lines. Doppler broadening and pressure broadening cause the absorption lines to be smeared and result in the wide oxygen absorption complex for low atmospheric layers. In addition to the line absorption, the non-resonant continuum absorption (i.e. the increasing absorption with frequency) of water vapor plays an important role. Cloud droplets emit depending on their mass extinction coefficient and show a quadratic increase of absorption with frequency (Fig. 4.1).

From ground-based microwave radiometers which operate in the two frequency bands (A, B, see Fig. 4.1) the following atmospheric parameters can be obtained: IWV principally from the strength of the absorption line, water vapor profiles from the pressure broadening along the absorption line A and LWP from the increasing absorption with increasing frequency, especially in the window frequencies around 31 GHz. Temperature profiles are retrieved from multi-channel measurements along the oxygen absorption complex B.

4.1.2 HATPRO system

As part of this thesis, a HATPRO microwave radiometer (Rose et al., 2005) was operated in Nangatchori. The field work included setup, calibration and maintenance of the instrument. After the data had been gathered, a detailed quality control was made. In order to derive atmospheric variables from the BTs, statistical retrieval algorithms for the climatic conditions in West Africa were developed. Quicklooks for BTs, IWV, LWP and profiles are available at http://gop.meteo.uni-koeln.de/~hatpro/amm/ for every day.

4 Instrumentation and measurement principles

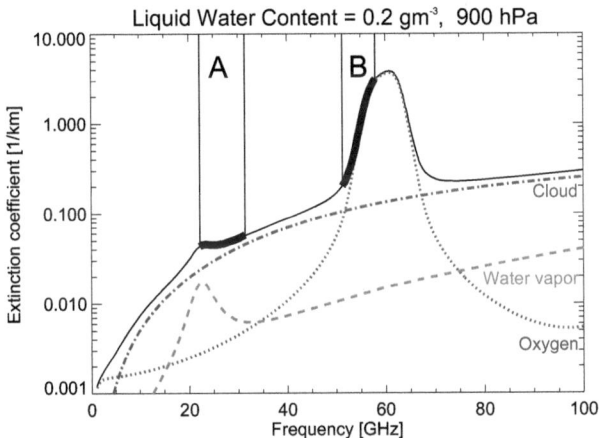

Figure 4.1: Atmospheric extinction in the microwave spectrum for a cloudy atmosphere. The red line shows the oxygen contribution, the green line the water vapor and the blue line the contribution of liquid water assuming a liquid water content (LWC) of $0.2\,\mathrm{gm}^{-3}$. The black line represents the sum of all components. Measurements by the HATPRO microwave radiometer are performed in the frequency bands A and B.

HATPRO continuously measures all 14 channels (Tab. 4.1) simultaneously. Observations are possible during nearly all weather conditions. Only when precipitation causes wet antennas or radomes no useful measurements can be performed. A high power dew blower is installed to remove water droplets on the radome. Auxiliary measurements of air pressure, temperature and humidity are automatically performed by HATPRO. A rain detector is used to flag periods of rain, and a GPS clock is used for exact time synchronization of the measurements.

Calibration errors are the most frequent source of radiometer inaccuracies. Therefore, HATPRO applies several calibration methods. Absolute calibration is performed by a cooled liquid nitrogen (LN_2) target and a temperature stabilized (hot) target inside the radiometer. Secondary calibration include a noise-diode which can be switched to the receiver inputs and a gain calibration. In addition, a tipping curve calibration is used for channels with a low opacity. This method is restricted to homogeneous, clear sky conditions. All calibration methods are performed automatically except the absolute calibration using LN_2 which has to be done manually. During the instrument deployment in Nangatchori, two LN_2 calibrations have been performed in January and June 2006, respectively. More details on the calibration methods can be found in Rose et al. (2005).

4.1 Microwave radiometry

HATPRO was operated in the zenith pointing mode most of the time (except when performing elevation scans, see below) giving a set of 14 BTs every 2 seconds. In order to obtain a consistent one-year data set, the zenith pointing observation mode was tilted by 20 degrees to the north at Nangatchori to avoid direct sunlight at any time of the year. As a consequence, the optical path through the atmosphere is increased by 6.4 % which has been taken into account within the retrieval algorithms. A 24-hour timeseries of BTs on 11 June 2006 is presented in Fig. 4.2. The frequencies along the water vapor line show a decreasing BT with increasing frequency, indicating the reduced absorption along the line. In the course of the day, a steady BT rise was observed. This increase was strongest in the 22.24 GHz channel and is thus related to an increasing water vapor content during the day. The peaks in the channels along the water vapor line are stronger for higher frequencies and therefore are caused by cloud liquid water. At the higher frequencies along the oxygen line (54–58 GHz) the atmosphere is optically dense. For example, the 58.0 GHz channel is so optically thick, that it represents the mean physical temperature within a layer of 300–400 m around the receiver. Therefore, the diurnal cycle of the BT at 58.0 GHz corresponds to a certain degree with the air temperature measured at the ground. With lower frequencies, the radiation systematically stems from higher layers and the diurnal cycle smears out.

Data Quality Checks

In order to get high quality data, erroneous measurements have to be filtered out. The first error source, namely precipitation, is detected automatically by a rain sensor. During these periods, water droplets on the radome contaminate the measurements. In the example of Fig. 4.2, a short rain shower was detected by the rain sensor after 20 UTC (green bar on top of the BT curves) which caused a cooling of the atmosphere. Besides an additional automatic check which rejects spikes that are beyond any possible atmospheric variability, a manual (or visual) quality check is performed to find all suspect data. During the rainy season in Africa some nights with dewfall had to be sorted out manually. The rain sensor did not detect dew and since the blower was not heated during this campaign, it was not possible to remove the dew droplets in case of 100 % relative humidity (fog). The importance of this quality check can be seen in Fig. 4.3 where after the rain event the LWP only slowly goes back to the zero line, indicating that dew droplets remained on the radome. Therefore, a manual filter (red bar) was set between 20 and 22 UTC.

Retrieval algorithm development

Statistical algorithms are used in order to retrieve atmospheric parameters (e.g., IWV, LWP, or temperature profiles) from BTs. The algorithms are developed on the ba-

4 Instrumentation and measurement principles

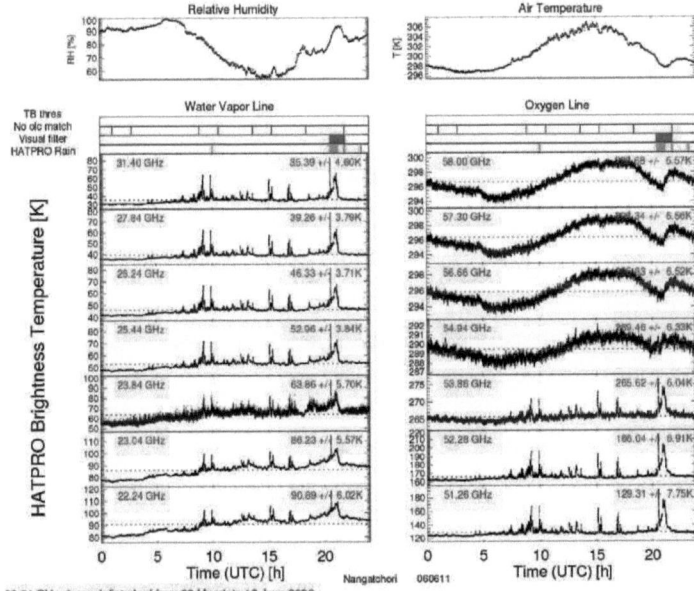

Figure 4.2: HATPRO BT on 11 June 2006 in Nangatchori. Top: Relative Humidity and Air Temperature measured by external sensor on the radiometer. Below: Quality flags ("TB thres": set if BT exceeds a valid range. "No olc match": set if data are only available for the water vapor frequencies. "Visual filter": set by manual inspection. "HATPRO rain": set if rain sensor on). Bottom: 24-hour timeseries of BT for the 14 HATPRO frequencies. Daily mean value and standard deviation shown for each frequency. Note that the 23.84 GHz channel was still disturbed by the MRR at that time (see section 4.3).

4.1 Microwave radiometry

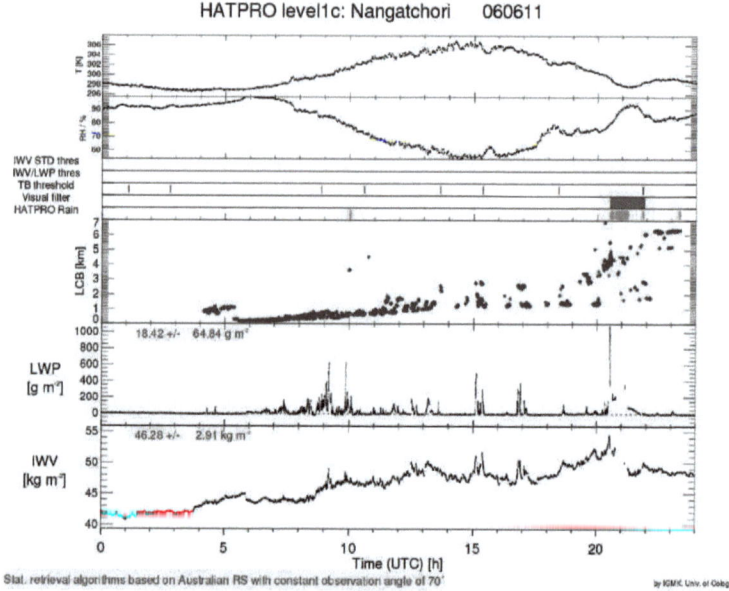

Figure 4.3: 24-hour timeseries of IWV and LWP from HATPRO observations in Nangatchori as well as the lowest cloud base (LCB) detected by the ceilometer on 11 June 2006.

sis of synthetic observations generated from a representative long-term radiosonde data set (Löhnert and Crewell, 2003). Figure 4.4 presents the steps which are taken in the algorithm development. From the input radiosonde data set (profiles of temperature $T(z)$, humidity $q(z)$ and pressure $p(z)$), the LWC is calculated by a cloud model using a modified adiabatic approach. The cloud boundaries are determined by a 95 % relative humidity threshold (Karstens et al., 1994). In the following, the radiosonde data set is split into two parts, a training and a test data set. From the training data set, a radiative transfer (RT) model calculates synthetic BTs that a microwave radiometer would observe at the ground. By means of a multi-linear regression, the statistical retrieval algorithm is derived. This algorithm is then applied to the test data set to assess the quality and determine the theoretical retrieval error.

It has to be kept in mind that these retrieval algorithms are limited for the range of atmospheric conditions for which they have been trained. Although radiosondes were

4 Instrumentation and measurement principles

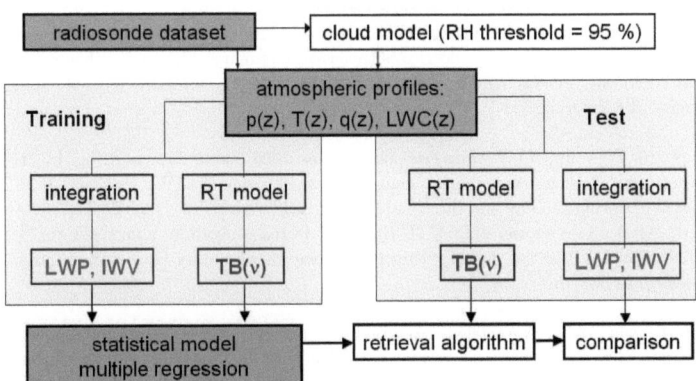

Figure 4.4: Flow chart for the development of statistical retrieval algorithms for ground-based microwave radiometers from a radiosonde data set using a RT model and mulitilinear regression.

launched from April to October 2002 as well as during most time of the year 2006 at Parakou, Benin (100 km east of Nangatchori), this data set is by far not large enough for developing a robust retrieval algorithm. Due to the lack of sufficient high quality radiosonde data in Western Africa all retrieved HATPRO products used in this study were generated by applying retrieval algorithms based on a 15-year data set of about 10000 radiosoundings from Darwin, Australia. Climatic conditions in Northern Australia with marked dry and wet seasons are similar to those in central Benin. This is confirmed by comparing mean profiles and frequency distributions of the Darwin soundings with those from Parakou. The only adaptation made was to compress the original radiosonde profiles by 385 m within the lowest 10000 m in order to account for the higher elevation of Nangatchori (415 m vs. 30 m) which otherwise would lead to a water vapor bias. Because all atmospheric states observed in Parakou are included within the Darwin statistics we are confident of the retrieval quality.

Coarse profiles of water vapor and temperature can be derived from the multi-spectral measurements when viewing in zenith direction. While the integrated water vapor (IWV) can be obtained with a high accuracy of less than $1\,\mathrm{kg\,m^{-2}}$ from the strength of the water vapor line, information on the vertical distribution is limited to two to four independent layers with an RMS uncertainty of $1-2\,\mathrm{g\,m^{-3}}$ (Löhnert et al., 2009). The reason for this is that the microwave frequencies along the 22.235 GHz water vapor absorption line are strongly correlated and thus do not provide more information on the vertical distribution of water vapor. For this reason, strong vertical humidity gradients cannot be reproduced

4.1 Microwave radiometry

properly with the microwave radiometer measurements. The accuracy of LWP retrievals is about $25\,g\,m^{-2}$ (Löhnert and Crewell, 2003) for two-channel radiometers. Adding more channels along the water vapor line decreases the error only marginally. A significant improvement could only be reached by adding higher frequency observations (e.g. 90 GHz, see Crewell and Löhnert (2003)).

An example for IWV and LWP timeseries on 11 June 2006 (same day as in Fig. 4.2) is presented in Fig. 4.3. The atmosphere is constantly moistening, the IWV increases from 41 to $54\,kg\,m^{-2}$ during the day and the cloud base is rising with the daytime development of the PBL. After a rain shower at 20 UTC the IWV decreases again to about $48\,kg\,m^{-2}$. The variability of LWP in the clouds within the developing boundary layer is quite large with values up to $600\,g\,m^{-2}$.

The vertical resolution of the retrieved temperature profiles from zenith pointing measurements is less than 2 km in the lowest 4 km (Liljegren et al., 2005) resulting in an RMS accuracy of 0.5 K near the surface and less than 1.6 K in the middle troposphere (Güldner and Spänkuch, 2001). The vertical resolution is limited through the weighting functions of the different frequency channels rather than through the radiometric noise. Additional measurements under six lower elevation angles were performed in order to enhance the accuracy and the vertical resolution of temperature profiles, assuming horizontal homogeneity of the atmosphere. When observing under lower elevation angle the observed radiation at relatively opaque frequencies along the oxygen line systematically originates from lower altitudes. Since these BTs vary only slightly with elevation angle, the method requires a highly sensitive and stable radiometer. Crewell and Löhnert (2007) could demonstrate HATPRO's ability to derive high-resolution temperature profiles of the PBL by comparisons with radiosondes and temperature measurements by a 100-m-mast. When using observations at six elevation angles between 5 and 90° the retrieval performance for the lowest 1500 m of the troposphere is significantly improved compared to zenith mode with an error of less than 0.5 K. The retrieval combines BTs measured at the four highest frequencies (54.94, 56.66, 57.30, and 58.00 GHz) under 5.4°, 10.2°, 19.2°, 30.2°, 42° and 90° elevation. The use of high (relatively opaque) frequencies limits the vertical information to about 1.5 km. The vertical distance of retrieved height levels is 50 m near the ground and gradually rising to 200 m in the highest levels. It should be noted that the true resolution might be worse especially for the higher levels depending on the actual situation, e.g. in case of multiple temperature inversion layers. The vertical resolution is best within a couple of 10 m close to the surface and gradually decreases to about 300 m at 400 m height (Westwater et al., 1999).

In Nangatchori, boundary layer scans were performed about every 10 minutes in order to observe the evolution of the lower PBL layers with high vertical as well as temporal resolution. Using the profiles of temperature and specific humidity, other meteorological variables to determine the atmospheric stability have been calculated, such as potential temperature, equivalent potential temperature and relative humidity. Measurement examples of high-resolution boundary profiles for four different times of the year are

4 Instrumentation and measurement principles

presented in Section 5.2 (Figs. 5.18 and 5.19). They give an impression of HATPRO's potential to monitor the PBL with high temporal as well as vertical resolution.

Since the two radiometers at Niamey did not perform elevation scans, only the IWV observations are used in this work. Temperature profiles derived from zenith observations alone are not very useful for boundary layer studies because shallow inversions close to the ground can only be detected by observations at low elevation angles.

4.2 Lidar Ceilometer

Two Vaisala CT25K Lidar Ceilometer instruments (Rogers et al., 1997) continuously operated in Nangatchori and Niamey during the entire year of 2006. The CT25K sends out light pulses at 905 nm and measures atmospheric backscatter with a temporal resolution of 15 seconds and a vertical resolution of 30 m (100 ft) in 250 range gates up to 7500 m (25000 ft) distance. To avoid direct sunlight during daytime the ceilometer was tilted by 20 degrees to the north (as the HATPRO microwave radiometer). This enabled us to observe throughout the whole year with the same angle even at June solstice when the sun is 15 degrees north of Nangatchori and in the same direction as HATPRO. The range correction for cloud base heights is done automatically by the instrument. In Niamey an identical instrument was used for the observation of clouds which was pointing vertically like the collocated microwave radiometers.

The basic purpose of the ceilometer is to detect cloud base heights. This is internally done by the Vaisala software using first a conversion of backscatter to extinction by Klett inversion (Klett, 1981) and a successive threshold to detect up to three cloud layers[1] simultaneously. As a further application, the backscatter profiles can indicate layers of enhanced aerosol content. This is possible especially during nighttime when no solar background reduces receiver sensitivity and the low vertical mixing in the boundary layer leads to distinct layers of different aerosol content.

A 24-hour timeseries of backscatter profiles for 11 June 2006 is presented in Fig. 4.5 (the same day as in Figs. 4.2 and 4.3). On that day many different atmospheric features can be detected. Between 00 and 04 UTC a distinct aerosol layer was present at about 2500 m AGL. After 04 UTC, this layer was mostly obscured by lower clouds (stratus) which cause strong scatter and quick attenuation of the signal. The extinction at the ceilometer wavelength is about 10 km^{-1} for dense water clouds compared to about 0.15 km^{-1} for clear atmosphere (Rogers et al., 1997). Slightly more than one hour after sunrise (0532 UTC), the base of the stratus clouds starts to rise, indicating that the convective boundary layer (CBL) starts to grow. In the afternoon the CBL reaches a depth of about 1400 m. After 20 UTC, clouds above 4000 m AGL appeared. These are likely ice clouds

---
[1] More than one cloud layer can be detected only if the lower ones are thin and semi-transparent.

4.3 Micro Rain Radar

because of the high vertical extension of backscattered signal. Water clouds would cause total attenuation of the signal more rapidly. At 21 UTC, a short rainshower is detected (orange-red signal between 0 and 1500 m AGL) which was also detected by the rain sensor of HATPRO (Fig. 4.3).

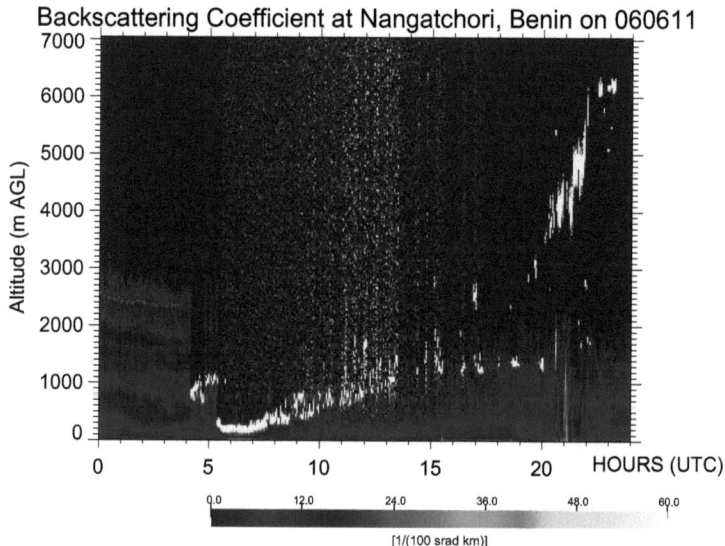

Figure 4.5: Ceilometer backscatter profiles over Nangatchori on 11 June 2006.

4.3 Micro Rain Radar

A micro rain radar (MRR) was operated in Nangatchori in 2006 in order to observe precipitation characteristics with high temporal resolution. The MRR is a low-power, vertical pointing frequency modulated continuous wave (FM-CW) Doppler radar, operating at a frequency of 24.1 GHz. It measures Doppler reflectivity $\eta(f)$, i.e. the spectral volume backscattering cross section at the Doppler shift f. From these Doppler spectra, profiles of drop size distribution $n(D)$ can be derived, using the relation between the terminal fall velocity of drops and their diameter D, following Atlas et al. (1973). On the basis of $n(D)$, profiles of radar reflectivity, liquid water and rain rate are calculated (Peters et al., 2002). The temporal resolution of the observations is 10 seconds. The

4 Instrumentation and measurement principles

range resolution is variable between 10 and 200 m, resulting in a maximum possible observation height between 300 and 6000 m AGL. As the operating frequency is higher than for standard C-Band weather radars, the attenuation at medium and high rain rates becomes noticeable and at large rain rates (> 50 mm/hour) already the second range gate is shadowed by the rain drops of the lowest gate. Attenuation is already important at rain rates of 10 mm/hour, which occur quite frequently in the tropical rain events. Therefore, the derivation of a rain rate is not straightforward and easily produces errors.

The MRR operated in Nangatchori from March to November 2006. The range resolution was initially set to 200 m, which results in a maximum range of 6000 m AGL, but was reduced to 160 m in June 2006 (maximum range then at 4800 m AGL). The resolution was changed because of the frequent obscurance of higher levels due to attenuation by heavy rain. The higher resolution provided more information within the low levels but with 4800 m range it was still possible to capture the melting layer (bright band) which was typically between 4000 and 4500 m AGL. Since the MRR and the passive microwave radiometer HATPRO operate in the same frequency range (K-Band), one channel of HATPRO (23.84 GHz, bandwidth 250 MHz) was influenced by the radiation emitted by the MRR (24.1 GHz) when these two instruments were located directly next to each other. For the second part of the campaign, the MRR was displaced by about 100 m on the backside of a small building to avoid any direct influence between the two instruments. After that, the HATPRO measurements were not disturbed anymore.

Three examples of rain events in 2006 are presented here to provide an overview over the instrument's capabilities:

On 9 September (Fig. 4.6), several isolated convection cells were present over the Central Benin area. To get an impression on the relation between radar reflectivity and rain rate, conventional rain gauge measurements next to the MRR measured 17.6 mm rainfall between 17 and 18 UTC and another 0.5 mm from 19 to 20 UTC. Because of the stratiform precipitation and the low rain rate at the latter event, the melting layer (bright band) can be seen clearly at about 4000 m AGL (no vertical motions mixed up the bright band).

On 28 July (Fig. 4.7), a squall line passed over Nangatchori, bringing 36.5 mm of rain, the largest intensity with 6.5 mm in 5 minutes occurred between 0600 and 0605 UTC.

The third example presents a case before the start of the monsoon season (Fig. 4.8). The days before 23 March 2006, moist air penetrated far north and a large mesoscale convective system (MCS) developed south of Nangatchori on 22 March. On the following day, deep convection developed over the area bringing after 15 February the second significant rainfall event for 2006 in Nangatchori. Due to the dry low-level air (and the very dry soil) at that time of the year, much of the falling precipitation after 15 UTC evaporated before it reached the ground.

4.3 Micro Rain Radar

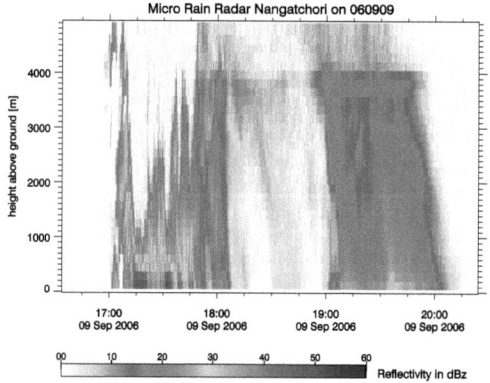

Figure 4.6: Radar reflectivity on 9 September 2006 as a function of height from MRR measurements.

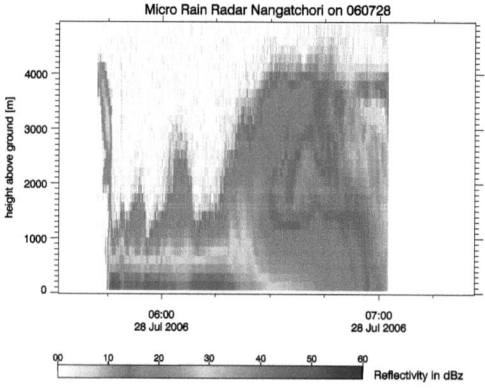

Figure 4.7: Same as 4.6, but for 28 July 2006.

4 Instrumentation and measurement principles

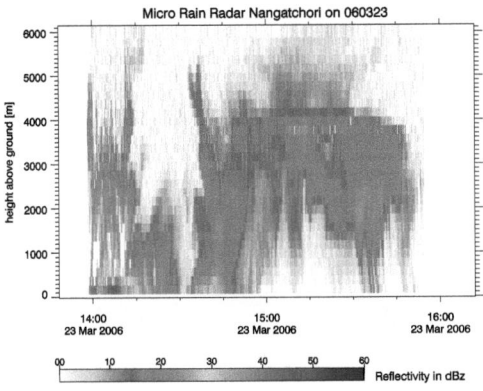

Figure 4.8: Same as 4.6, but for 23 March 2006.

4.4 Additional instrumentation

The UHF (1.24 GHz) wind profiler Doppler radar, operated by the Centre National de Recherches Météorologiques, provides vertical profiles of the three wind components, reflectivity and spectral width under clear air or precipitation events in the lowest 3 to 5 km of the atmosphere. It is a five static antenna profiler working in a continuous operating mode with a maximum peak power of 4 kW. The typical vertical and temporal resolutions are 100 m and 5 minutes, respectively. For more details on the instrument and data processing refer to Jacoby-Coaly et al. (2002) or Heo et al. (2003). This instrument was used to determine horizontal winds around the ITD and to validate the model results in chapter 6.

During AMMA, a network for the monitoring of IWV by using the Global Positioning System was established in West Africa (Bock et al., 2008). The atmospheric water vapor content can be obtained by determining the amount of water vapor causing the delay of the GPS signal between the different satellites and the receiver on the ground (slant water). IWV can be considered as the average over all these observations scaled to zenith, and over a certain period of time. GPS receivers were installed at six different locations in 2005 within the AMMA area. In order to get long-term water vapor observations, the observations continued after the main AMMA campaign (LOP). In this study, the data from the two sites in Djougou and Niamey are used. The accuracy of GPS IWV measurements is on the order of $0.5 \, \text{kg} \, \text{m}^{-2}$ (Bock et al., 2008). It has to be noted that the IWV by GPS represents the mean IWV value of a volume between the different

4.4 Additional instrumentation

satellite slant paths.

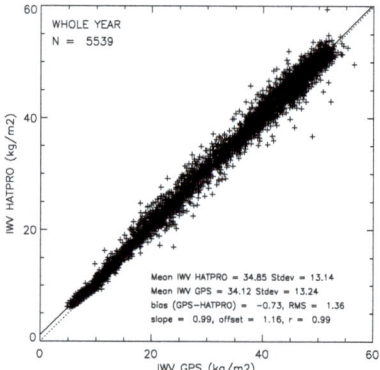

Figure 4.9: 1-hour mean IWV GPS vs. HATPRO microwave radiometer (whole year 2006).

Thanks to the one year period where both HATPRO microwave radiometer and GPS were continuously observing water vapor in the Djougou area, it was possible to perform a statistical comparison of IWV measured by these two methods. The GPS receiver was located at Djougou, being 10 km away from Nangatchori (location of HATPRO) at the same MSL height. For a total of 5539 hourly mean values, the HATPRO derived IWV exhibits a positive bias of $0.73 \, \text{kg} \, \text{m}^{-2}$ and a RMS difference of $1.36 \, \text{kg} \, \text{m}^{-2}$ for the year 2006 (Fig. 4.9). The same bias appears both during dry and wet seasons (Fig. 5.13) which shows that this difference has no seasonal dependency.

Several possible sources may contribute to this bias: first, in the retrieval of IWV from HATPRO data, uncertainties in the water vapor absorption model can lead to an error. Another uncertainty lies in the processing of raw GPS data or more precisely in the determination of the hydrostatic delay of the GPS signal, i.e. the contribution from the dry atmosphere. One method is to derive the hydrostatic delay from numerical weather prediction (NWP) models before the IWV can be calculated. This process is not straightforward, and is explained in detail by Bock et al. (2008). As the bias is rather independent from the season, it is more likely that the processing of GPS data contributes more to this error. On the contrary, if the HATPRO retrieval is biased, a seasonal dependency is more likely.

Meteosat Second Generation (MSG, Schmetz et al. (2002)) SEVIRI (Spinning Enhanced Visible and Infrared Imager) observations are used as a proxy for surface temperature in Nangatchori. As the continuum absorption of water vapor is relatively weak at the wave-

4 Instrumentation and measurement principles

length of 10.8 μm and the surface/cloud emissivity is close to the emissivity of a black body, the BTs measured in this channel are highly correlated to the target temperatures. For this reason, this channel is suitable for the detection of clouds due to their temperature which is generally lower than the temperature of the surface underneath. For the same reason clear sky BTs can be used as a proxy for the surface skin temperature. Over the study region, the spatial resolution of the observations is particularly high (3 km) due to the proximity to the satellite nadir at $0\,°N$, $0\,°E$. The temporal resolution for MSG products is 15 minutes.

4.5 Data availability

It was a special challenge to operate high-tech instruments continuously over one full year's cycle. Climatic conditions (heat, very dry air in dry season, high humidity in monsoon season), dust, insects and unreliable power supply made it difficult to compile an uninterrupted data set. Especially at the Nangatchori site, power cuts were frequent and thus, uninterruptable power supplies and back-up batteries were installed to ensure an autonomous operation of up to three hours. In case of longer power cuts, reliable procedures for automatic restarting of the whole system were arranged.

With all these preventative measures, it was possible to obtain a satisfactory data availability of more than 75 % for the whole year of 2006 for the main instruments presented in this work. The HATPRO microwave radiometer (section 4.1) operated 76 % of the time. The Lidar ceilometer (section 4.2) was even more reliable with a data availability of 87 %. The lower availability of HATPRO was caused by a broken axis of an internal mirror on 31 January 2006 which had to be replaced. Therefore, the instrument was not operating until 22 March 2006 when the spare part was brought in from Europe. The MRR operated only from 22 March to 2 November because no rainfall was expected before and after these dates. During the rainy season, the data availability of the MRR was only limited by power failures.

An overview of the HATPRO availability is given in Fig. 4.10 which shows that power cuts were more frequent during the wet season (May-October). Tables .1 to .4 in the appendix present the data availability as well as the weather situation for all days between June and September 2006 (SOP 1–3).

For the Niamey site, conditions were better as the instruments were directly located at the international airport of the capital, where a technician was permanently available and power supply was much more reliable. At this site, however, the instruments suffered more from the extensive heat and the very dry and dusty conditions (frequently over 40 °C and less than 10 % relative humidity).

4.5 Data availability

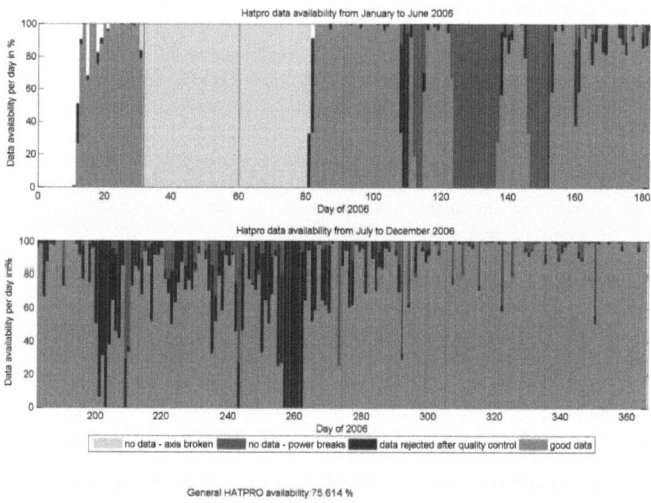

Figure 4.10: HATPRO data availability in 2006.

5 Analysis of the annual and diurnal cycle of water cycle parameters

The atmosphere over tropical West Africa is characterized by a very marked difference between the dry north-easterly Harmattan flow originating in the Sahara desert and the south-westerly monsoon flow from the Atlantic Ocean (Gulf of Guinea). Following the path of the sun, the ITD starts to move northward from the Guinea Coast in March, reaching its northernmost position in July and August at about 20 °N. The intensive observations by ground-based remote sensing instruments which were performed at the two locations of Nangatchori and Niamey (400 km north of Djougou) in 2006 made it possible to monitor the atmospheric state over a full year's cycle of the WAM climate.

The goal of this section is to explore this unique data set to achieve a new view on characteristic conditions of various atmospheric parameters, both over an annual cycle and for short timescales. The main emphasis lies on water vapor, cloud cover, cloud liquid water, temperature and humidity profiles, their behavior at different time scales (annual, seasonal and diurnal) and interdependencies between some of these parameters. Since in the tropical regions inter-diurnal fluctuations of the atmospheric state are much smaller than in mid-latitudes, monthly or bi-monthly averages provide a good overview over the atmospheric state in the course of the year. Thanks to the high temporal resolution of the observations, it is also possible to describe the diurnal cycle of the lower atmospheric layers.

5.1 Annual cycle

The most pronounced feature in the annual cycle is the variability of the atmospheric water vapor load. Several characteristics appear when regarding Fig. 5.1 which illustrates the annual cycle of IWV for Nangatchori and Niamey in 2006. The dry season in the first months of the year (January to April) was unusually moist over the region and characterized by several outbreaks of humid monsoon air from the south at both locations. In Djougou a major rain event on 15 February brought more than 50 mm of rain with a daily mean IWV of $42\,\mathrm{kg\,m^{-2}}$ which is close to monsoon season values of about $48\,\mathrm{kg\,m^{-2}}$. The strong northward moisture transport doubled the IWV even at Niamey which is 400 km to the north of Nangatchori. Within two days after the rain event at

43

5.1 Annual cycle

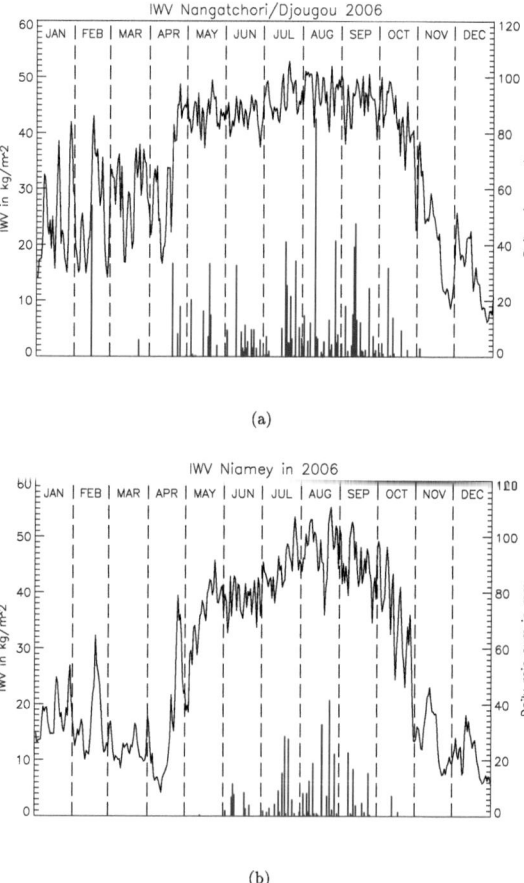

Figure 5.1: IWV annual cycle (black) and daily rain accumulations (blue bars) in 2006. IWV daily mean values derived from microwave radiometer measurements. Data gaps are filled with GPS observations. a) Nangatchori/Djougou. b) Niamey.

5 Analysis of the annual and diurnal cycle of water cycle parameters

Nangatchori, the daily mean IWV in Niamey reached $32\,\mathrm{kg\,m^{-2}}$ on 16 February. The driest conditions were present in early April, particularly at Niamey with IWV values of 3–$5\,\mathrm{kg\,m^{-2}}$, which is also a typical value for polar regions. Towards the end of April, the atmosphere moistened quickly. The IWV remained rather constant throughout the whole rainy season in Djougou with an average IWV of $44.0 \pm 5.2\,\mathrm{kg\,m^{-2}}$. However, the strongest precipitation activity which produced 62 % of the annual rainfall was connected to the monsoon peak from 15 July to 30 September when the atmosphere was slightly moister than before (IWV = $46.6 \pm 3.9\,\mathrm{kg\,m^{-2}}$).

This second moistening phase is in line with the description of the monsoon onset by Sultan and Janicot (2003). The ITCZ is situated quasi-stationary at 5 °N in May and June and moves to a second quasi-stationary position at 10 °N in July and August. This abrupt shift is called monsoon onset. It is caused by the meridional land-sea contrast and is characterized by a temporary decrease of convection over the whole of West Africa. For the Djougou area the term "monsoon onset" is somewhat misleading as already in May and June quite strong rainfall events took place. Following Sultan and Janicot (2003), the average date of monsoon onset is 24 June. On 3 July, the transition phase with the installation of the monsoon and convection over the Sahel is completed. In 2006, the transition period with weakened convection lasted longer (25 June to 10 July) and the monsoon was only fully established by 20 July 2006 (Janicot et al., 2008; Drobinski et al., 2009).

At Niamey, two distinct phases of moistening can be distinguished as well (Fig. 5.1(b)). The first phase called pre-onset starts about the same time as in Djougou with a moist air surge at the end of April. Another dry spell followed in early May. In the end of May and in June, IWV values of $40\,\mathrm{kg\,m^{-2}}$ are reached. Some slight rainfall events already occurred in June, but the major moistening took place in the second half of July which agrees with the findings of Janicot et al. (2008). The southward retreat of the monsoon air started in early October (Niamey) and took place much faster than the onset. The last rainfall in Niamey was recorded on 19 October, and already two weeks later on 2 November Nangatchori received its last rainfall. In December, the atmosphere was very dry with several days of IWV values below $10\,\mathrm{kg\,m^{-2}}$. Unlike the year before, in January 2007 very dry conditions and strong Harmattan winds predominated over the whole region.

The higher water vapor variability at Nangatchori during the dry season is also presented in the cumulative IWV frequency distribution in Fig. 5.2. During the wet season with rather constant conditions (May–October 2006), 90 % of the IWV observations range between 35 and $50\,\mathrm{kg\,m^{-2}}$, whereas in the dry season (January–April and November–December 2006), the variation is much larger (90 % of the observed values lie between 8 and $41\,\mathrm{kg\,m^{-2}}$)

The amount of rainfall which is a very critical and variable parameter for the Sahelian climate (Le Barbé et al., 2002) follows the cycle of IWV (Fig. 5.1). In Nangatchori, the

5.1 Annual cycle

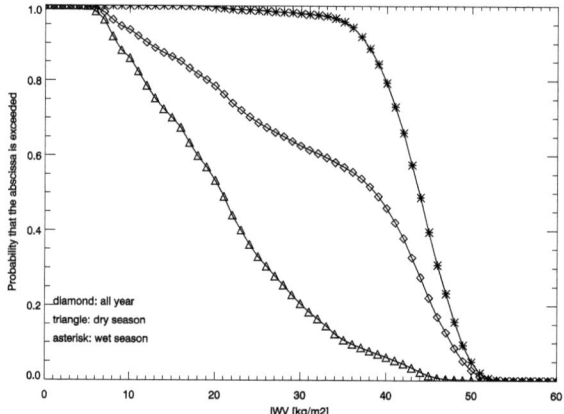

Figure 5.2: Cumulative frequency distribution of IWV from HATPRO measurements at Nangatchori. Dry season: January–April and November–December 2006. Wet season: May–October 2006.

dry season in early 2006 was interrupted by two rain events on 15 February and 23 March. From the end of April rainfall became more regular, but decreased slightly in June with a dry phase in early July connected to the ITCZ shift. From mid July to the end of September frequent rain events occurred, mainly connected with MCS's and squall lines. October brought a quick decline in rainfall. The total precipitation resulted in 1124 mm which was below the average of 1340 mm (1950–70) and 1208 mm (1970–90), respectively. Niamey received 384 mm of rain in 2006 which was also well below the long-term mean (1905–1989) of 564 mm (Lebel et al., 1997). First considerable precipitation occurred in early June, but most of the rain fell between mid July and the end of September. The last rainfall of the season was observed on 19 October. The monthly rain sums for both locations can be found in Tab. 5.4, along with the amount of cloud cover.

The sharp changes of water vapor and rainfall conditions make it possible to define distinct seasons. Following Sultan and Janicot (2000, 2003), Lothon et al. (2008) defined the onset of the monsoon by analyzing the zonal wind component and the water vapor mixing ratio. They identified four distinct periods (dry, moistening, wet, and drying, see Tab. 5.1). Following their definition, the moistening period starts with the occurrence of positive zonal winds (monsoon flow). The dry season in early 2006 showed frequent and pronounced moist air outbreaks which were associated with southwesterly winds and therefore the moistening period starts on 1 January. Only about four weeks in December 2005 matched their dry season criterion with no positive zonal winds.

5 Analysis of the annual and diurnal cycle of water cycle parameters

Table 5.1: Seasons in 2006 as defined by Lothon et al. (2008).

	Djougou	Niamey
dry	–	1 Jan–20 Apr
moistening	1 Jan–25 Apr	20 Apr–15 Jul
wet	25 Apr–27 Oct	15 Jul–1 Oct
drying	27 Oct–10 Nov	1 Oct–5 Nov

Table 5.2: Sub-periods of the monsoon season in 2006 as defined by Bock et al. (2008).

	Description	Djougou	Niamey
A	installation of humid air mass	10–25 Apr	1–20 May
B	stationary period	25 Apr–20 Jun	20 May–3 Jul
C	further IWV increase at monsoon onset	20 Jun–20 Jul	3–27 Jul
D1	core of rainy season	20 Jul–10 Sep	27 Jul–10 Sep
D2	start of monsoon retreat	10 Sep–10 Oct	10 Sep–1 Oct
E	monsoon retreat, last rainfall events	10 Oct–5 Nov	1 Oct–2 Nov

In order to identify specific subperiods of the monsoon season, Bock et al. (2008) made a more precise analysis of the rainfall and water vapor conditions during the wet season of 2006 (Tab. 5.2). The "installation of the humid air mass" period was much shorter than the "moistening" defined by Lothon et al. (2008). It started about 20 days earlier in Djougou than in Niamey, and was characterized by similar mean IWV values (Tab. 5.3). Period B was moister in Djougou than in Niamey, indicating a longer duration of the establishment of the humid air mass in Niamey. IWV in periods C, D1 and D2 were quite similar at both locations. Niamey shows a clear maximum in D1 ($48.0\,\mathrm{kg\,m^{-2}}$), being even moister by $1.3\,\mathrm{kg\,m^{-2}}$ than Djougou ($46.7\,\mathrm{kg\,m^{-2}}$). However, it has to be noted that the MSL height of Niamey is only 223 m whereas Djougou lies 415 m above sea level.[1] The IWV variance (see Tab. 5.3) is by far largest in periods A and E which proves the large fluctuations caused by the sharp water vapor gradients that are situated over the area at that time. During the peak monsoon season, relatively steady moisture conditions were observed.

The annual cycle of cloud cover (Fig. 5.3) reflects the annual distribution of IWV and rainfall quite well. Cloud cover was observed by lidar ceilometers at Niamey and Nangatchori. At both locations, the cloud cover maximum can be found in July and August which is connected with the precipitation maximum and a minimum diurnal temperature range (Fig. 5.4). Clouds in dry season are mainly associated with mid-level humidity (see later this section). For the whole year of 2006, 22.7 % of the time was cloudy in Nangatchori, and only 9.2 % in Niamey (Tab. 5.4). During the core of the rainy season

---

[1] Assuming an atmospheric water vapor content of $20\,\mathrm{g\,m^{-3}}$ (which is a typical value in the monsoon season close to the ground) and similar conditions aloft, a IWV difference of up to $3.8\,\mathrm{kg\,m^{-2}}$ can be explained simply by the height difference.

5.1 Annual cycle

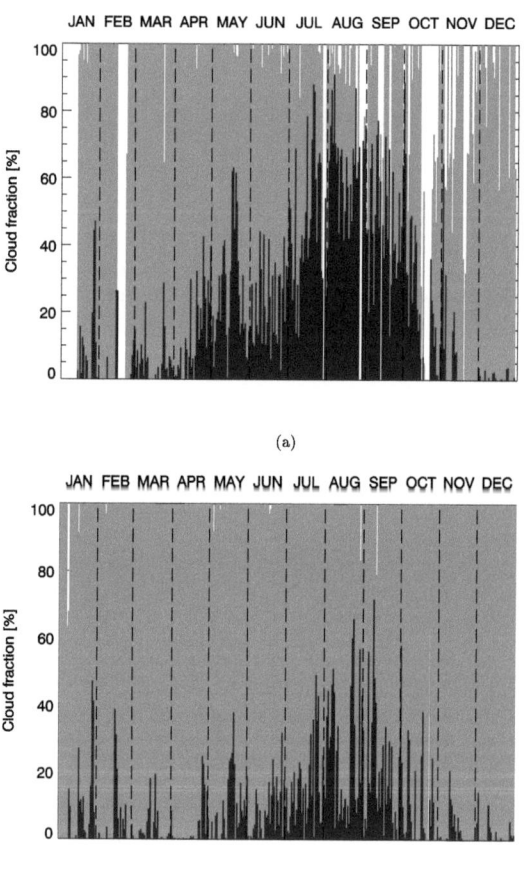

Figure 5.3: Daily percentage of cloudy scenes from ceilometer observations: a) Nangatchori/Djougou, b) Niamey. Black: percentage of day with clouds detected by ceilometer, grey: percentage of day without clouds, white: no data.

5 Analysis of the annual and diurnal cycle of water cycle parameters

Table 5.3: Mean IWV and standard deviation for different parts of the monsoon season 2006, defined in Tab. 5.2.

	Djougou	Niamey
A	$33.5 \pm 10.3\,\mathrm{kg\,m^{-2}}$	$31.9 \pm 7.3\,\mathrm{kg\,m^{-2}}$
B	$43.1 \pm 3.6\,\mathrm{kg\,m^{-2}}$	$39.7 \pm 4.7\,\mathrm{kg\,m^{-2}}$
C	$44.5 \pm 3.9\,\mathrm{kg\,m^{-2}}$	$44.5 \pm 4.8\,\mathrm{kg\,m^{-2}}$
D1	$46.7 \pm 4.0\,\mathrm{kg\,m^{-2}}$	$48.0 \pm 6.7\,\mathrm{kg\,m^{-2}}$
D2	$46.3 \pm 3.7\,\mathrm{kg\,m^{-2}}$	$44.6 \pm 4.9\,\mathrm{kg\,m^{-2}}$
E	$38.2 \pm 6.8\,\mathrm{kg\,m^{-2}}$	$33.0 \pm 10.1\,\mathrm{kg\,m^{-2}}$

Table 5.4: Monthly percentage of cloudy times observed by ceilometer (below 7000 m AGL) and rainfall (in mm) from rain gauge observations in Djougou and Niamey.

	Djougou clouds	Niamey clouds	Djougou rain	Niamey rain
Jan	9.8 %	6.9 %	0.0	0.0
Feb	3.6 %	4.6 %	51.8	0.0
Mar	5.2 %	2.8 %	3.4	0.0
Apr	13.1 %	3.4 %	51.2	0.0
May	25.9 %	8.6 %	139.2	0.5
Jun	23.2 %	8.8 %	76.9	43.1
Jul	47.3 %	15.9 %	236.3	102.0
Aug	64.3 %	25.9 %	227.4	160.7
Sep	47.7 %	19.9 %	246.2	67.0
Oct	25.7 %	10.2 %	88.8	8.8
Nov	6.5 %	2.4 %	3.4	0.0
Dec	0.8 %	1.6 %	0.0	0.0
year	22.7 %	9.2 %	1124.2	382.1

(August), the difference is even more striking (64 % cloudy times in Djougou and only 26 % clouds in Niamey). It has to be kept in mind that thin cirrus and all clouds with bases higher than 7000 m AGL were not detected by the ceilometer. Although the atmospheric moisture content (IWV, see Fig. 5.1) is very similar for both locations during the peak monsoon season in August, the much lower cloud content is caused by several mechanisms. With the shorter rainy season, the amount of annual rainfall is much lower in Niamey. Hence, the growing period is shorter and the vegetation is less dense. This results in a lower capability of water storage through plants which in turn leads to lower evapotranspiration and a lower moisture supply for the low atmosphere. In addition, higher temperatures in Niamey compared to Djougou during monsoon season (Fig. 5.4) reduce the number of days with fog considerably.

The annual temperature range in the tropics is much smaller than in mid-latitudes. Nevertheless, when observing Fig. 5.4 in detail, many features of the WAM system and

5.1 Annual cycle

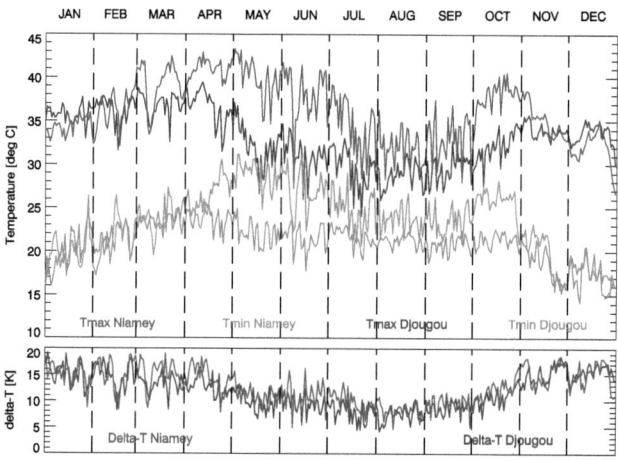

Figure 5.4: Annual cycle of daily maximum (T_{max}) and minimum (T_{min}) temperature measured in 2 meters above ground at Djougou and Niamey in 2006 (top) as well as diurnal temperature range (T_{max} minus T_{min}, bottom).

the general circulation can be recognized. In the dry season (January and February, as well as November and December), temperatures in Niamey and Djougou are quite similar with maximum values around 35 °C and minimum values around 20 °C. Drier periods are associated with a stronger diurnal cycle (up to 18 K), whereas under moist conditions the daily temperature range remains below 10 K. From March on, Niamey starts to heat up considerably stronger than Djougou. In April and May, temperatures in Niamey reach 40 °C and more and also during night temperatures rarely drop below 30 °C. In May, Nangatchori is cooler by almost 10 K, but the diurnal temperature range is about the same at both locations.

The temperature maximum in Niamey is in May. It is caused by the position of the sun which is in zenith at that time with a maximum of incoming solar radiation. Nangatchori would receive virtually the same amount of solar flux, but this region further south is already influenced by the monsoon air mass. High humidity and frequent cloud cover reduce the conversion of absorbed solar radiation into sensible heat at the ground. After the monsoon onset in Niamey in early July, temperatures decrease there as well. The diurnal range between maximum and minimum becomes smaller, reaching its minimum values in August which is the month with the highest IWV and highest cloud cover at both

5 Analysis of the annual and diurnal cycle of water cycle parameters

locations (Fig. 5.1). In October, a secondary temperature maximum occurs in Niamey after the monsoon retreat. There is no such maximum in Djougou, as the area is still more influenced by the humid air mass. In November and December the solar radiation is already considerably lower and the temperatures remain below 35 °C with nighttime minima around 15 °C which is much cooler than in January 2006 when Harmattan winds were much weaker than average.

The interdependence of diurnal temperature range and cloud cover is illustrated in Fig. 5.5. Both diurnal maximum and minimum temperatures are higher, respectively lower if no clouds influence the radiation balance. On cloudless days within a dry air mass, nighttime minima are usually lower because of the large negative radiation budget at the surface. On the other hand, the maximum temperature is governed by the amount of solar radiation which reaches the ground to be converted into sensible heat. Therefore, all cloudless days show a difference between maximum and minimum temperature of at least 10 K in Nangatchori with an extreme value of 18 K. The cloudiest days (with cloud cover of 60 % or more) are associated with a diurnal temperature range of about 5 K. In addition, all days but one having a temperature range of less than 10 K show a diurnal mean IWV of at least 36 kg m^{-2} in Nangatchori. When considering only days with cloudy periods of more than 5 %, the correlation between cloud cover and temperature range is acceptable for Nangatchori (0.71), but much poorer for Niamey (0.44) which might be due to the low number of sample days with a considerable cloud cover.

Apart from the cloud cover, the diurnal temperature range also depends on the atmospheric moisture. High relative humidity during the monsoon season is responsible for nighttime dewfall and fog formation. Moreover, water vapor is an efficient greenhouse gas. With a high atmospheric water vapor content, the radiation balance at the ground changes significantly and the temperature usually does not drop as low. Therefore, a good correlation (0.82) between the daily temperature range and the average daily long-wave radiation balance at the ground was found for Niamey (Fig. 5.6). Unlike that, the diurnal temperature range depends only weakly on the mean daytime shortwave radiation balance (Fig. 5.7).

Days with strongly negative long-wave radiation balance show a large difference between daily minimum and maximum temperature. The nighttime minimum temperature in 2 m AGL is highly connected to the long-wave radiation balance as the nocturnal cooling under windless conditions is mainly determined by the radiation balance. The downward long-wave radiation flux is mainly determined by clouds (resp. their base temperature) and the atmospheric water vapor content, whereas the upward long-wave radiation flux depends on the surface temperature. In the dry season, nearly all incoming solar radiation is converted into sensible heat which contributes to an increase of surface temperature. During monsoon season when the soil is wet, a large portion of incoming radiation is used for evaporation and the surface temperature cannot rise as high. Due to this latent heat conversion, and due to the higher water vapor content in the atmosphere, the long-wave radiation balance is less negative in the wet season even during cloud-free periods.

5.1 Annual cycle

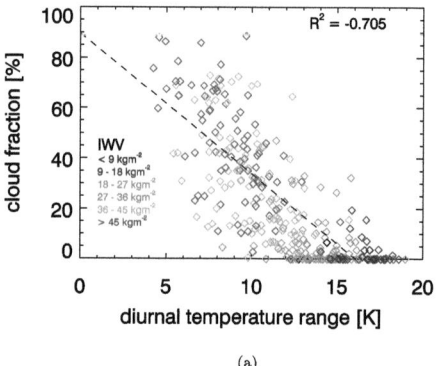

(a)

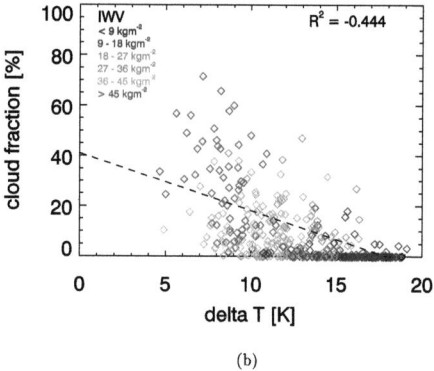

(b)

Figure 5.5: Diurnal 2m-temperature range (T_{max} minus T_{min}) as a function of daily cloud cover for all 2006. a) Nangatchori/Djougou. b) Niamey. Colors indicate the diurnal mean IWV value.

5 Analysis of the annual and diurnal cycle of water cycle parameters

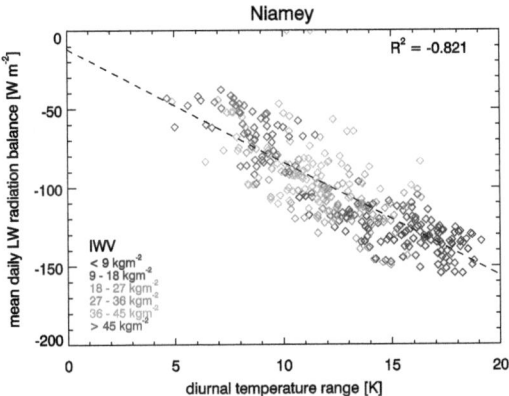

Figure 5.6: Diurnal 2m-temperature range (T_{max} minus T_{min}) as a function of daily mean long-wave radiation balance in Niamey for the whole year 2006. Colors indicate the diurnal mean IWV value.

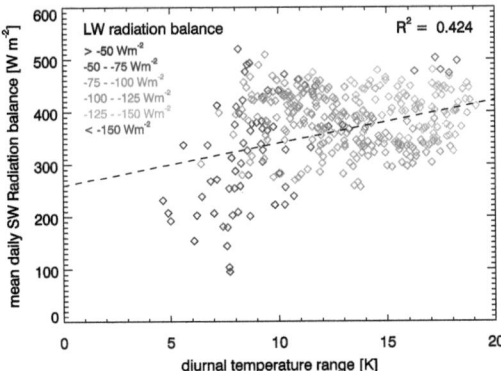

Figure 5.7: Diurnal 2m-temperature range (T_{max} minus T_{min}) as a function of shortwave radiation balance (averaged over daytime between sunrise and sunset) in Niamey for the whole year 2006. Colors indicate the diurnal mean long-wave radiation balance.

5.1 Annual cycle

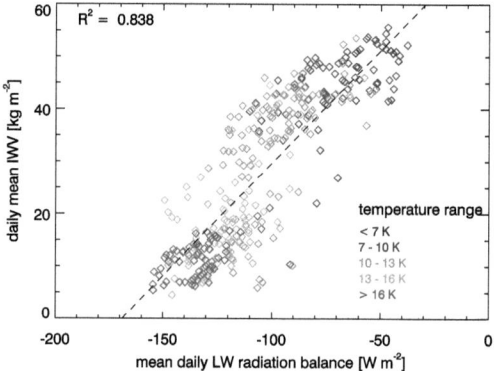

Figure 5.8: Daily mean long-wave radiation balance as a function of daily mean IWV in Niamey for the whole year 2006. Colors indicate the diurnal temperature range.

Therefore the diurnal temperature in August and September is smallest over the whole year (Fig. 5.4). As the dependence between mean daily shortwave radiation balance and temperature range is only weak, one may conclude that the daily temperature range at Niamey is mainly driven by the long-wave radiation balance which in turn is determined by the IWV rather than by the incoming solar radiation.

The good correlation between long-wave radiation balance and daily mean IWV (Fig. 5.8) is also caused by the low amount of cloud cover over Niamey. For other regions of the earth (e.g. mid-latitudes), this correlation is very poor as clouds occur more frequently and often independently from the IWV. Unfortunately, no reliable data for long-wave radiation were available for Nangatchori where clouds were more frequent than in Niamey.

To summarize, the annual cycles of IWV, daily temperature range and radiation balance are presented in Fig. 5.9. The long-wave radiation balance at the surface is negative for the whole year, i.e. the daily mean long-wave radiation flux is always pointing upwards with a minimum in August and a maximum in early April. At that time, the sun reaches its first zenith position (which is indicated by a maximum in solar radiation) and the air mass is still very dry. The second zenith position of the sun is in August at the peak monsoon season. On some cloud-free days, the mean shortwave radiation balance reaches similar values as in April, but due to the high water vapor content, the long-wave balance is much less negative and the diurnal temperature range is much lower.

5 Analysis of the annual and diurnal cycle of water cycle parameters

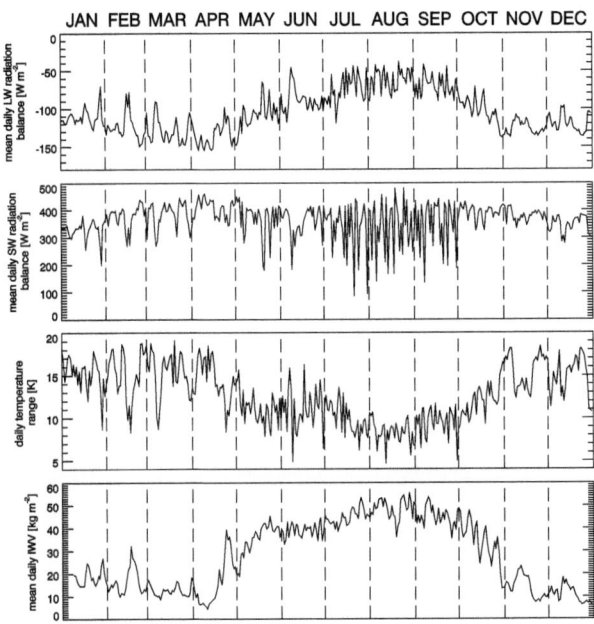

Figure 5.9: Annual cycle of daily mean long-wave radiation balance, diurnal temperature range and daily mean IWV for Niamey in 2006.

The examination of the annual cycle of various atmospheric parameters over Niamey showed that the main driver for changes is the water in the atmosphere, both as vapor and as clouds. As the area is close to the equator, the annual fluctuation of incoming solar radiation is only small. At Nangatchori, also the cloud cover appeared to be an important factor on temperature and radiation balance, especially during the monsoon season.

5.2 Diurnal cycle

The previous section presented the annual variation of the conditions in the lower atmosphere over West Africa which is mainly controlled by the annual monsoon cycle and the related strong water vapor differences between dry and wet season. The focus of this section will lie on the variability of the diurnal cycle in the lower atmosphere with regard to the monsoon cycle. The instrumentation deployed in the area in 2006 made it possible to study the diurnal cycle of the lower atmosphere with a high temporal resolution for the first time in West Africa. Before the AMMA campaign started, such a detailed examination was not possible because no suitable observations were available.

First of all, the mean diurnal cycle of potential temperature for different months is presented in Fig. 5.10. During the wet period (May–September), the difference between monthly mean temperature maximum and minimum is much lower (5 K) than during the dry season (up to 10 K) because of the less negative radiation balance during the monsoon season caused by clouds and water vapor. Therefore, a closer look will be taken on the diurnal cycle of these parameters.

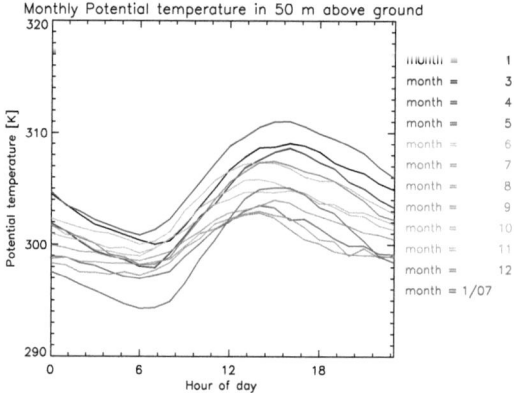

Figure 5.10: Monthly mean diurnal cycle of potential temperature in 50 m AGL over Nangatchori from microwave radiometer observations.

The diurnal distribution of cloud cover is an indicator for the humidity in the lower atmosphere. The averaged occurrence of clouds over Nangatchori and Djougou is presented in Fig. 5.11 for six two-month-periods. During all of these periods, a distinct diurnal cycle of cloud base heights can be recognized. However, the height evolution of daytime convective clouds differs considerably between the seasons. This is particularly true for

5 Analysis of the annual and diurnal cycle of water cycle parameters

Nangatchori (Fig. 5.11(a)), whereas in Niamey (Fig. 5.11(b)) clouds on top of the PBL were present nearly only during the months of July and August.

For this reason, a more detailed analysis of the annual cycle of the PBL is discussed for Nangatchori only. Between January and April, a few days with daytime convection occurred. These days are connected with IWV values higher than average, i.e. moist air surges from the south (see Fig. 5.1). As the soil is generally very dry without green vegetation at that time, evaporation is small and the mixed layer can grow fast, reaching a depth of about 2 km on these cloudy days. Note that during cloud-free days with lower IWV, a PBL depth of more than 3 km was observed by HATPRO. In May and June, PBL clouds become more frequent, reaching a base height of 1200–1400 m in the afternoon. At the peak of the monsoon season (July/August), clouds are frequently observed on top of the boundary layer which does not grow deeper than 800–1000 m at that time of the year. The different PBL depth between May/June and July/August cannot be explained by the atmospheric water vapor, as these values are quite similar throughout all of these months (see Fig. 5.1). An important factor for the lower PBL depth is the vegetation which normally starts to establish in May and June and is fully developed only by July. The energy and moisture balance at the ground changes significantly with more dense vegetation. In contrast, the vegetation in Niamey remains rather sparse even during monsoon season with only scattered trees and some grass.

The onset of convection and the growth of the mixed layer starts between 8 and 9 UTC during all seasons. Due to the proximity to the equator, daytime is quite constant throughout the year. Note that in Djougou sunrise varies between 0531 and 0615 UTC, whereas sunset is between 1728 and 1818 UTC. For Niamey, sunrise times are between 0524 and 0619 UTC and sunset takes place between 1721 and 1823 UTC.

At both sites, the cloud base height distribution reveals a second significant cloud maximum at 4–5 km AGL. These clouds are associated with the African Easterly Jet (AEJ) and can be seen throughout the whole year, more frequently occurring during the wet season. Due to the lower number of days with masking PBL clouds in Niamey they contribute much larger to the total cloud occurrence in Niamey (69 %) than in Nangatchori (28 %, see Fig. 5.11). This different cloud distribution can also be seen in diurnal rainfall patterns between June and September[2] (Fig. 5.12). In the Djougou area, the rainfall maximum is in the afternoon hours with a minimum at the morning. For Niamey, the conditions are quite different. Most of the rainfall occurs during nighttime and in the morning hours with a minimum in the early afternoon. The rainfall in the Niamey area results from westward-moving disturbance lines with a maximum during the early morning hours. In contrast, Djougou shows an afternoon maximum which is related mainly to local thunderstorms during the peak monsoon season (see Buckle (1996)).

---

[2]This restriction was made because high-resolution rain gauge data for Nangatchori are only available for this period. However, a large portion of the annual rainfall occurs during this period.

5.2 Diurnal cycle

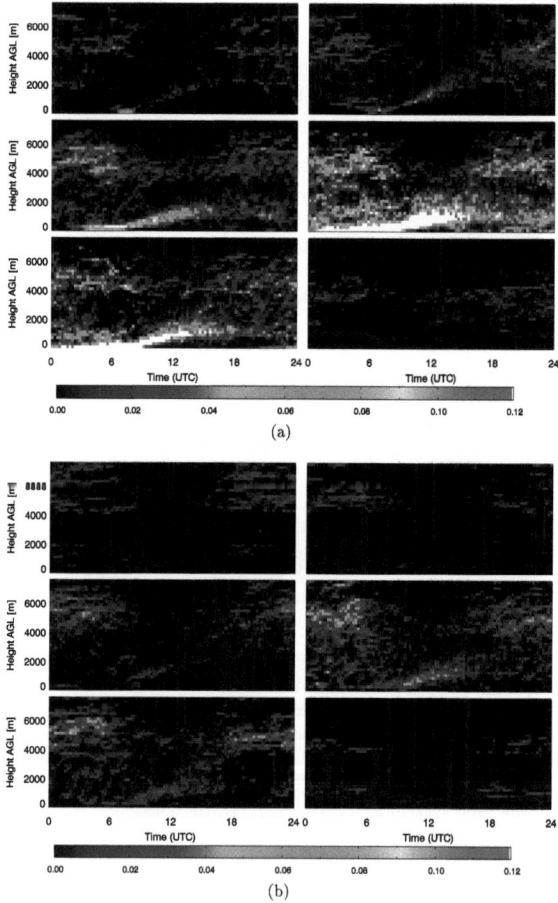

Figure 5.11: Cloud base height frequency distributions from ceilometer observations for two months' periods at a) Nangatchori and b) Niamey. Top left: Jan/Feb, top right: Mar/Apr, center left: May/Jun, center right: Jul/Aug, bottom left: Sep/Oct, bottom right: Nov/Dec.

5 Analysis of the annual and diurnal cycle of water cycle parameters

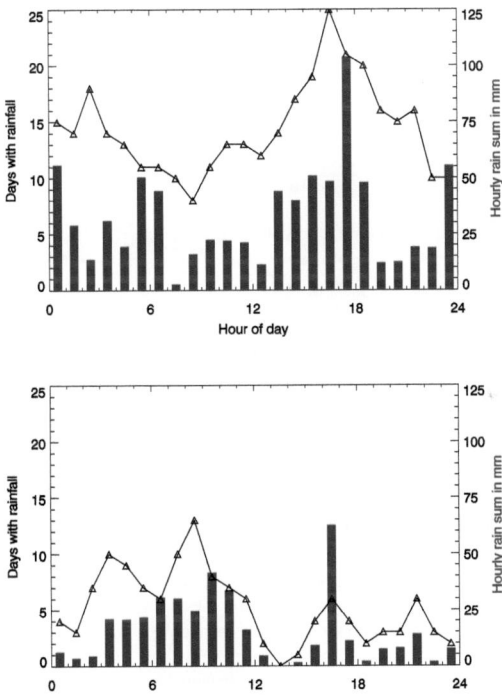

Figure 5.12: Diurnal cycle of rainfall for the period between 1 June and 15 September 2006. Black line represents the number of days where rainfall was observed during a certain hour. Blue bars show cumulated rainfall over the whole period. Top: Nangatchori, Bottom: Niamey.

5.2 Diurnal cycle

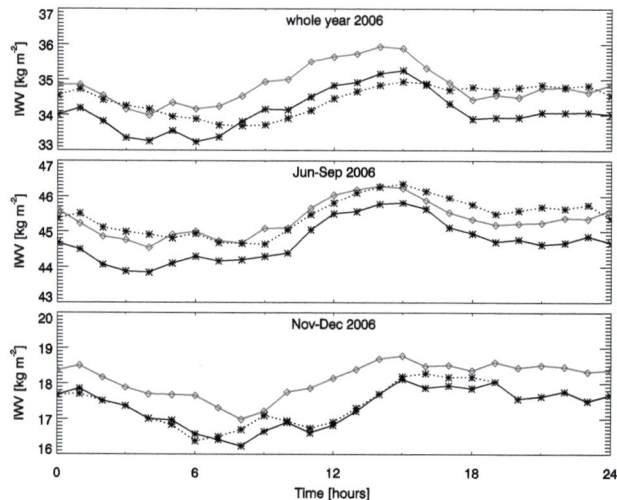

Figure 5.13: Mean diurnal cycle of hourly mean IWV observed in Nangatchori/Djougou for different periods in 2006. Red: HATPRO, black: GPS. Dashed line includes also GPS measurements from times where no HATPRO measurements are available.

The diurnal cycles of IWV and LWP are much less pronounced than for rainfall. For Nangatchori, Fig. 5.13 shows that during all seasons of the year, a weak diurnal cycle was observed $(1-2\,\mathrm{kg\,m^{-2}})$. The IWV minimum occurs in the early morning hours. During daytime, the IWV increases, reaching a maximum at about 15 UTC. In the evening and until midnight, the IWV again starts to decrease slowly. The IWV increase during daytime results mainly from evaporation at the ground (soil and vegetation). During nighttime, dewfall removes parts of this water vapor from the atmosphere again.

When comparing GPS and HATPRO, a bias of about $0.7\,\mathrm{kg\,m^{-2}}$ appears which has been found already in Fig. 4.9. The different behavior of HATPRO and GPS in the evening can be explained by frequent rainfall at that time (see Fig. 5.12) when HATPRO measurements were filtered out. It is likely that rainfall occurs with IWV values higher than average which are not taken into account and therefore, the average becomes lower.

The mean HATPRO derived LWP diurnal cycle in June–September 2006 at Nangatchori (Fig. 5.14) exhibits a distinct LWP maximum between 13 and 15 UTC which fits well with the growing mixed layer at that time. However, it has to be kept in mind that

5 Analysis of the annual and diurnal cycle of water cycle parameters

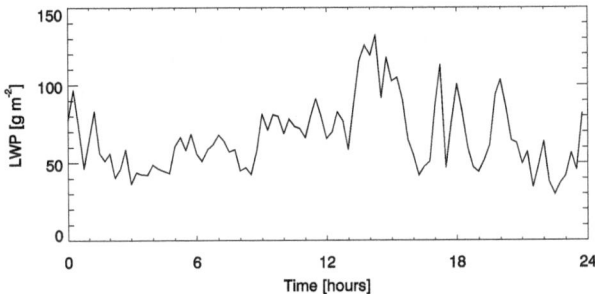

Figure 5.14: Mean diurnal cycle of LWP from HATPRO observations in Nangatchori (June–September 2006). Only cloudy times detected by the ceilometer were considered.

during rainfall events and in case of a wet radome, no reasonable observations can be taken by HATPRO and therefore an underestimation of LWP is very likely after 15 UTC like for IWV in Fig. 5.13.

Generally, only the lowest cloud base height can be derived from ceilometer observations (except in case of rather thin cloud layers). The combination of the lowest cloud base height with LWP observed by HATPRO is presented in Fig. 5.15. This frequency distribution gives some impression of the vertical distribution of cloud liquid water. Again, a bimodal distribution of cloud occurrence for both locations appears which can also be found in Fig. 5.11. However, lower clouds are more frequent at Nangatchori with a maximum at the ground (fog), whereas in Niamey mid-level clouds dominate over the year (maximum in about 5 km AGL). Closer examination reveals some further details of clouds over the area. In Nangatchori, LWP values of less than 200 gm^{-2} are mainly in connection with low clouds (in large parts nighttime fog). Higher LWP of 400 gm^{-2} and more shows a maximum at clouds with bases around 1000 m which is exactly the height of the PBL clouds mainly observed in July/August (Fig. 5.11(a)). The reason for this can be found in the growth of convective clouds in the course of the day. As more and more water condenses, the LWP value rises steadily until the maximum mixed layer depth is reached in the afternoon, if no precipitation occurs (Fig. 5.14).

An important factor for boundary layer processes is the atmospheric stability which is determined by the vertical gradients of (potential) temperature and humidity. From microwave radiometer measurements, it is possible to derive boundary layer temperature profiles and therefore to diagnose atmospheric stability. Therefore, the diurnal cycle of monthly mean temperature gradients derived from HATPRO observations is presented

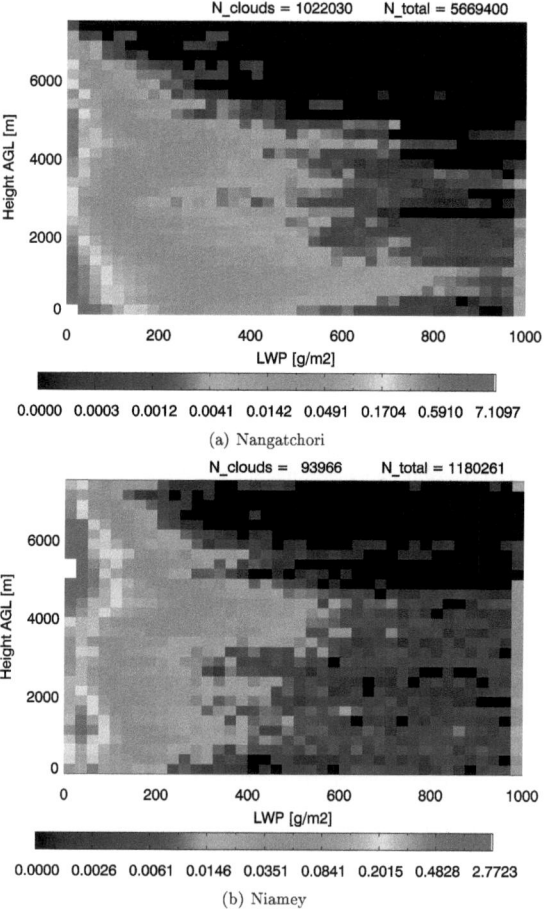

Figure 5.15: LWP versus cloud base height for 2006: a) Nangatchori, b) Niamey. For each bin, colors indicate the frequency of clouds relative to all measurements in each layer. Cloud base height is determined by ceilometers at both sites. LWP was observed by HATPRO in Nangatchori and by the two-channel radiometer in Niamey.

5 Analysis of the annual and diurnal cycle of water cycle parameters

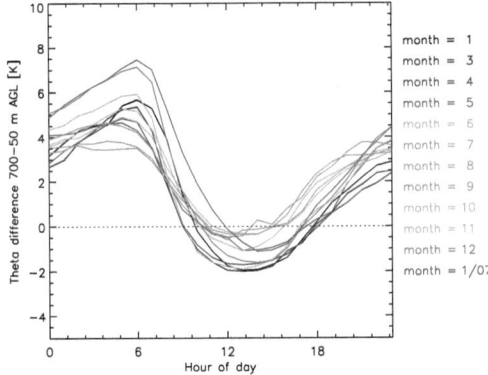

Figure 5.16: Diurnal cycle of potential temperature gradient between 700 and 50 m AGL from HATPRO observations for different months in 2006 and January 2007 in Nangatchori.

in Fig. 5.16. The two levels (700 m and 50 m AGL) were chosen because 50 m represents the lowest retrieved temperature level and 700 m is still within the PBL all year round. Positive values of the temperature gradient correspond to a (dry) stable atmosphere and negative values are observed under unstable conditions. During the dry season, strong nighttime inversions are frequent. Therefore, the mean monthly inversion strength at sunrise (around 06 UTC) is up to 7 K (December 06, January 07). On the other hand, negative values are seen during daytime. These super-adiabatic conditions show the substantial instability in lower levels during the dry season. The monsoon season is characterized by less strong nighttime inversions and potential temperature gradients around 0 during daytime.

From Fig. 5.16, also the big difference between January 2006 and January 2007 is striking. In 2007, the lower atmosphere was much more stable, both during day and night. Also the absolut values of temperature were considerably lower by about 5 K (Fig. 5.10). The constant advection of cool and dry air masses from the north and the total absence of moist tropical air lead to this large inter-annual difference. Longer-term observations would be necessary to evaluate the inter-annual fluctuations of the north-easterly Harmattan flow which has its typical maximum and southernmost extent in January.

The diurnal and annual cycle of stability and its vertical distribution can be derived from the profiles of potential temperature in Fig. 5.17. In the annual range it turns up that the highest daytime PBL temperatures occur in April. The sun has reached its zenith

5.2 Diurnal cycle

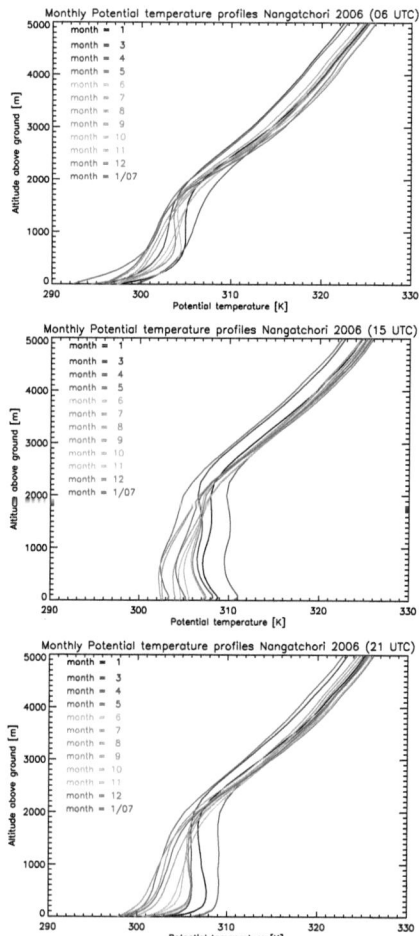

Figure 5.17: Mean monthly profiles of potential temperature. Top: 06 UTC, Center: 15 UTC, Bottom: 21 UTC.

5 Analysis of the annual and diurnal cycle of water cycle parameters

position over Nangatchori at that time, but the moist air masses have not yet reached the area. Therefore, a large portion of the incoming solar radiation is used to heat the PBL. The lowest daytime temperatures within the PBL were observed in the wet season (June-September). Frequent cloud cover prevents the sun to heat the ground and the PBL. The mean monthly PBL top can be seen best at 15 UTC. It is diagnosed as the level where the potential temperature gradient with height turns positive (at about 800 m AGL in July/August and 2200 m AGL in January to April). At 21 UTC (approximately 3 hours after sunset), at all seasons a shallow temperature inversion close to the ground was observed. The layer with constant potential temperature aloft (also called residual layer) is deepest in April, whereas during the monsoon season, the potential temperature gradient is positive throughout the whole PBL depth. In the morning (06 UTC), the lower atmosphere is stably stratified during all seasons. A remaining residual layer can be seen in November and December at about 1000 m AGL. The strongest temperature inversion close to the ground was observed during the dry months when radiative cooling of the lowest air layers is more effective than during the monsoon season with moist and cloudy air masses (see also Fig. 5.16).

Measurement examples of high-resolution boundary profiles over Nangatchori for special days of the year are presented in Figs. 5.18 and 5.19. For all these profiles, only data from HATPRO (including boundary layer scans) were used. Contrary to all the other figures in this chapter where diurnal cycles were presented before, these profiles are centered at midnight to particularly allow observing nighttime boundary layer conditions. On 12/13 April, the ITD was situated over Nangatchori, resulting in a strong diurnal water vapor cycle. The following date in early monsoon season (10/11 June) was characterized by a developing PBL and a shallow nighttime temperature inversion. On 8/9 August (during peak monsoon), very moist conditions with a rain shower at 16 UTC were observed. The fourth example (14/15 December) presents a typical day in the middle of the dry season.

Temperature and potential temperature profiles show the highest values in April, whereas the strongest nighttime inversion occurred in December. A less pronounced diurnal temperature cycle is observed in June and August. Relative humidity is very low in December, and especially high in August after a short rain event which took place at 16 UTC on 8 August. During nighttime, fog was observed (between 0 and 6 UTC, the relative humidity was 100 %). In April, a strong jump in all variables around midnight indicates the passage of the ITD and the arrival of moist air in Djougou (see also Chapter 6).

5.2 Diurnal cycle

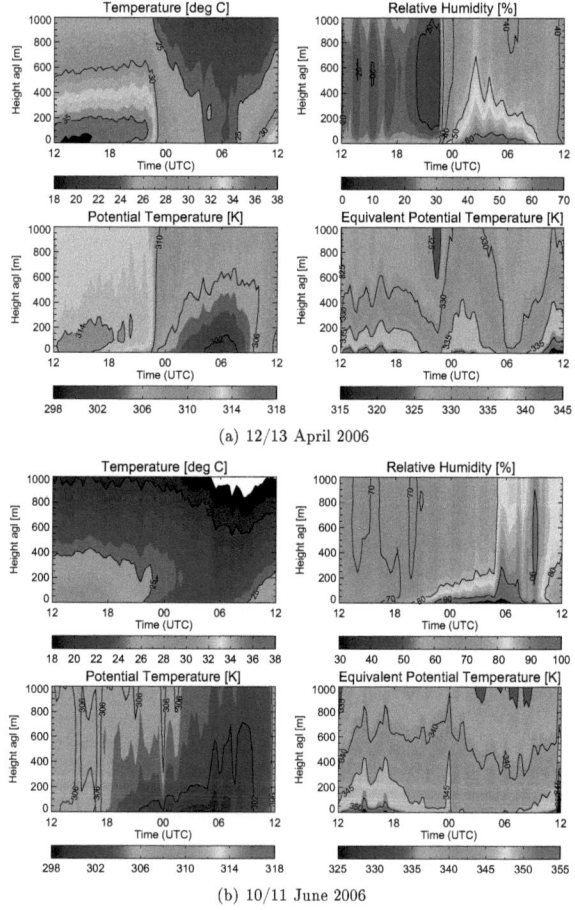

Figure 5.18: Diurnal cycle of time-height cross sections of temperature, relative humidity, potential temperature and equivalent potential temperature for a) 12/13 April 2006 and b) 10/11 June 2006. Note that the range of all values changes between the events.

5 Analysis of the annual and diurnal cycle of water cycle parameters

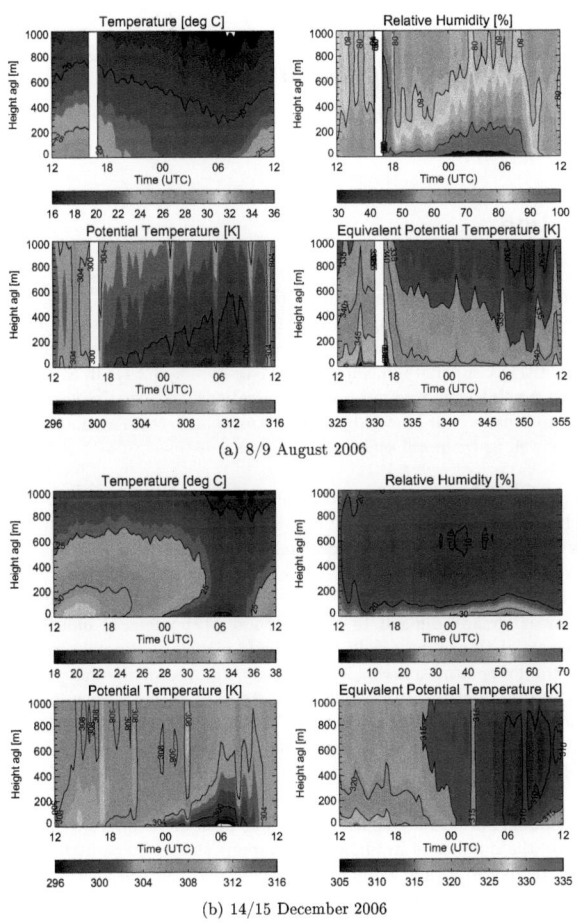

Figure 5.19: Same as Fig. 5.18, but for a) 8/9 August 2006 and b) 14/15 December 2006.

6 Detailed investigation of the ITD diurnal cycle over Nangatchori

6.1 Introduction

This section is focused on the diurnal cycle of the ITD position over Nangatchori in April 2006. The continuous monitoring of the ITD using a variety of ground-based and satellite remote sensing methods is presented here. The ground-based measurements were taken at the AMMA supersite in Nangatchori (see section 3.2.1). In a further step, the extension from a point view (ground-based observations) to a spatial view (satellite, model) will allow to broaden our knowledge of the atmospheric state around the ITD on different spatial as well as temporal scales.

The diurnal variation of the latitudinal ITD position during the study period is about 100–200 km. This is relatively small to be represented in global NWP models with a typical grid box size of 25–50 km. In operational mode, these models usually provide outputs only every three to six hours which is not adequate for monitoring this phenomenon properly. Moreover, the sparse network of routine ground-based and upper-air observations makes it already difficult to capture these phenomena in model analyses. Therefore, it is essential to use mesoscale atmospheric models to describe sub-synoptic scale features in the lower atmosphere. In this study, the use of the mesoscale model Méso-NH with a resolution of 10 x 10 km will demonstrate to what extent it is able to reproduce the atmospheric state and sharp gradients along the ITD by validating the model with observations. In addition, we will use the model as a tool to further investigate the processes around the ITD.

The choice of a suitable period for this case study has been made after considering two main prerequisites: the number of available observations should be as high as possible and the strongest temperature and humidity gradients between the moist and dry air masses should be situated over Nangatchori. For these reasons the period between 9 April, 12 UTC and 13 April, 00 UTC when the ITD remained close to Nangatchori was chosen for the model comparison.

First of all, a model description can be found in section 6.2 and the synoptic situation is presented briefly in section 6.3. Once the simulations have been validated, the full 3D

6 Detailed investigation of the ITD diurnal cycle over Nangatchori

information of the model is used to investigate the ITD development in section 6.4.

6.2 Mesoscale simulation

6.2.1 Model description

We use the non hydrostatic meso-scale model Méso-NH (Lafore et al., 1998) which has been developed jointly by CNRM/Météo France and Laboratoire d'Aérologie. Méso-NH contains a variety of different sets of parameterizations. For this study, turbulence (Bougeault and Lacarrère, 1989), convection (Kain and Fritsch, 1993; Bechtold et al., 2001) and biosphere-atmosphere thermodynamic exchanges (Masson et al., 2003) schemes are used. Méso-NH is coupled to an externalized surface model which handles heat and water vapor fluxes between the low-level atmosphere and four types of surface: vegetation, towns, oceans and lakes. The vegetation types are provided by the ECOCLIMAP database (Masson et al., 2003). Natural land surfaces are described by interactions treated in the Soil Biosphere and Atmosphere model (ISBA) (Noilhan and Mahfouf, 1996) in which the type of soil discretization is based on the force-restore method with 3 layers. Moreover, we used the radiative scheme of the European Centre for Medium-range Forecasts (ECMWF) which computes radiative fluxes of shortwave and long-wave radiation.

6.2.2 Simulation description

In this study, a simulation over 84 hours (between 9 April 2006 12 UTC and 13 April 2006 00 UTC) - was carried out over a domain of 1000 x 1000 km with horizontal mesh size of 10 km, centered at 9 °N and 2 °E (Fig. 6.1).

In the vertical, 72 stretched levels were used. The lowermost level was set at 10 m above the ground and 31 levels were located within the lowest 1000 m AGL, while the highest level was set at 28 km AGL. The model outputs are available every hour. In the first six hours, some spin-up problems might be evident.

Initial and lateral boundary conditions were taken from the ECMWF analyses of 0.5° horizontal resolution. In order to keep the simulation close to the analysed meteorological conditions, the boundary conditions of the simulations were nudged towards the ECMWF analyses with a 6-hour time scale. ECMWF temperature, humidity and wind fields are nudged throughout the depth of the atmospheric column. However, the observation data we use for the validation of the model were not assimilated into the ECMWF analyses (within the model domain only 83 sondes from Niamey and 23 radiosondes from Cotonou

6.3 Synoptic situation

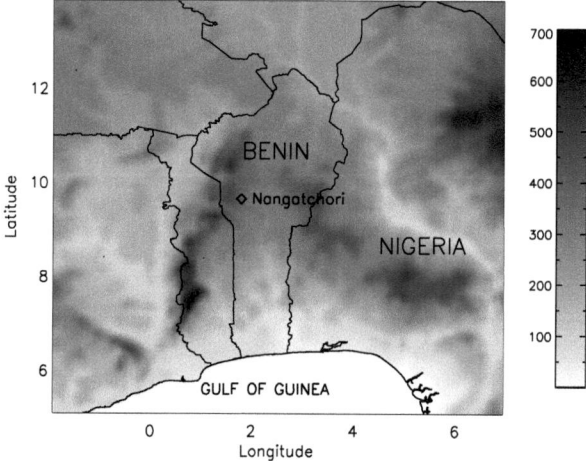

Figure 6.1: Méso-NH Model domain with orography (shaded). Djougou is situated 10 km northwest of Nangatchori.

were assimilated into the ECMWF analyses during the whole month of April (AMMA-EU, 2007)). Due to the nudging, possible changes induced by the presence of clouds are taken into account indirectly via the analyses. The validity of this assumption is confirmed by remote sensing observations which show that in the center of the domain (over Nangatchori) clouds were present only during three hours out of 84 of the model run. These clouds were thin broken altocumulus clouds at about 4 km AGL. The comparison of Méso-NH results with ECMWF analyses confirms the benefits of a mesoscale model when simulating processes around the ITD. We found much more pronounced gradients of water vapor, temperature and pressure across the ITD, a stronger low-level jet (925 hPa) as well as a deeper heat low in the afternoon (not shown here).

6.3 Synoptic situation

The dry season in early 2006 in the Djougou area was characterized by several outbreaks of moist air from south, resulting in moister than average conditions from January to

6 Detailed investigation of the ITD diurnal cycle over Nangatchori

March, and causing also a major rainfall event (>50 mm) on 15 February 2006 over the region. For that reason, it is not easy to define a clear beginning date of the transition towards the monsoon season. However, the region was predominantly influenced by a north-easterly flow until the end of March when winds started to shift every night from north-east to south-west, evidencing the daily northward pulse of the monsoon air. This nighttime wind shift was observed during 16 nights between 1 and 18 April 2006 with a mean shift time at 2319 UTC (see section 6.6). After 20 April, the study area remained all day in the south-westerly monsoon flow. Janicot et al. (2008) also stated that around 10 April dry air masses with very weak convective activity prevailed over West Africa, followed by abrupt increase in moisture towards the end of that month. As these diurnal pulsations have not been observed at moisture outbreaks before the end of March, they turned out to be an important precursor to the long-term establishment of moist air in the region.

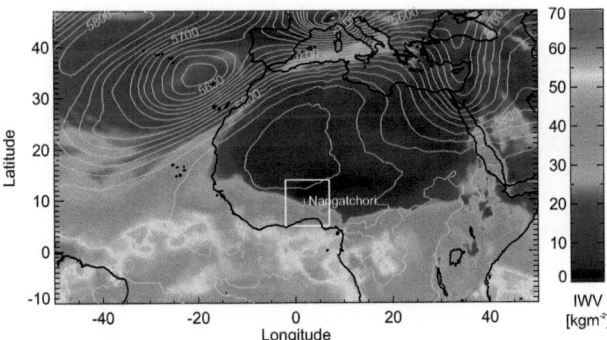

Figure 6.2: Synoptic situation on 11 April 2006, 00 UTC from ECMWF analysis. Integrated water vapor (colored shading) and 500 hPa geopotential height (contour lines with 20 gpm distance) are shown. White box shows the boundaries of the Méso-NH domain. Djougou is situated 10 km northwest of Nangatchori.

The development which led to the dry air outbreak arriving in the Djougou area on 10 April 2006 started several days before. At the beginning of April, the synoptic situation was characterized by a ridge over north-western Africa and a very pronounced mid-latitude trough over Libya and Egypt into Sudan. Over the western parts of the Sahara, high pressure was present. From 4 April on, the ridge was flattened and shifted eastwards

6.3 Synoptic situation

due to a rather strong westerly flow over Algeria, causing a lower surface pressure over the Sahara which induced a northward move of the ITD west of the 5 °E meridian. At the rear edge of another trough over the Red Sea, dry and relatively cool upper-tropospheric air subsided on 7 April over Libya and Northern Chad. Following a re-strengthening of the ridge over Algeria, these very dry air masses moved south-westward at the southern edge of the high geopotential over the central Sahara. The driest air masses arrived over the study area on 11 April with an IWV of less than $10\,\mathrm{kg\,m^{-2}}$. The boundary between the moist and the dry air masses was located at the Djougou latitude at 00 UTC (Fig. 6.2). On 10 and 11 April, the synoptic situation is characterized by a strong surface high pressure area between Libya and Eastern Niger which causes a southward retreat of the ITD east of 0° E whereas the heat low was placed over Mauritania. The ECMWF analysis of 925 hPa wind from 11 April 00 UTC (Fig. 6.3) illustrates the quite strong winds over southern Niger and northern Nigeria and the convergence zone between Harmattan and monsoon flow over central Benin (around 9 °N) very clearly. After 11 April, the north-easterly wind decreased due to the weakening of the high pressure area over Libya. As a result, the diurnal mean ITD position slowly moved back northward.

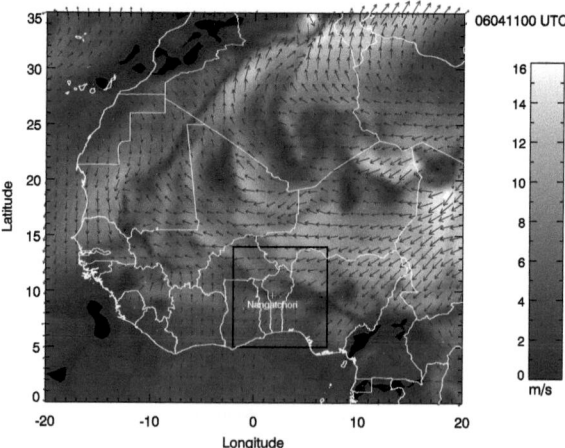

Figure 6.3: ECMWF analysis of 925 hPa wind speed on 11 April 2006, 00 UTC (grey shading). Superposed are wind arrows at 925 hPa. The resolution of the analysis is 0.5° x 0.5°. The black box marks the Méso-NH domain. The convergence zone within the study zone appears as a dark line of weak wind between the two flows.

6 Detailed investigation of the ITD diurnal cycle over Nangatchori

6.4 Dynamics and thermodynamics of the ITD

6.4.1 Surface observations

In this section we start with evaluating the model results using surface observations. In a further step, integrated water vapor as well as profiles of temperature, humidity and wind will be considered. We are aware of the problems that may arise when comparing point observations with a single model grid box. However, the period was mostly cloud-free which reduces sub-grid variability substantially. Furthermore, the vegetation is relatively uniform around Nangatchori and the area is quite flat (mean grid box heights differ by maximum 30 m over 10 km). The center of the closest model grid cell is 2 km away from Nangatchori at an altitude of 393 m, the real elevation of Nangatchori being 415 m. Observed differences in atmospheric conditions between Djougou and Nangatchori (10 km east-west distance) are small compared to the variability observed in the entire domain.

Temperature and dewpoint observations at 2 m height show a pronounced diurnal cycle (Fig. 6.4) with a consistent picture between Nangatchori and Djougou. In the afternoon, temperatures usually reach 40 °C at that time of the year. After sunset (1803 UTC), the temperature drops quickly, particularly in Nangatchori. This measurement site lies in a flat terrain outside a small village, surrounded by manioc fields and some shrubbery. In this open area, nighttime temperatures inversions are stronger than in the town Djougou where the measurement site is in an urban area on a small hill. In order to illustrate the dominant role of surface energy exchange we also consider the temperature of the lowest atmospheric layer (0–25 m) derived from HATPRO observations (see also section 4.1). The temperature values of the HATPRO profile were obtained by performing elevation scans with the microwave profiler. The temperature difference between this layer and the 2 m measurements in Nangatchori is up to 8 K before midnight, demonstrating the extremely strong temperature inversion close to the ground. Evidence for that is also given by temperature observations on a small tower in Nangatchori where the temperature difference between 1 and 4 m AGL is 5 K (Pospichal and Crewell, 2007). During daytime, due to strong surface heating, the observations close to the ground are warmer than the model, whereas during the first part of the night (strong inversion) the observations are mostly cooler. During all three nights of the simulation, the ITD moves from south to north over Nangatchori around 00 UTC and the atmosphere close to the ground mixes due to the strong southerly nocturnal jet. Moreover, it can be confirmed that the timing of the ITD passage does not differ by more than a few minutes between Nangatchori and Djougou (Fig. 6.4). Prior to the front arrival, the dry and warm Harmattan air was still present on top of the very shallow inversion (10 m). Just before midnight, the temperature at 2 m rises whereas the mean temperature between 0 and 25 m drops. Here, the 10 m model temperature corresponds roughly to an average between the remote sensing measurements in the 0–25 m range and the 1.2 and 2 m in-situ measurements.

6.4 Dynamics and thermodynamics of the ITD

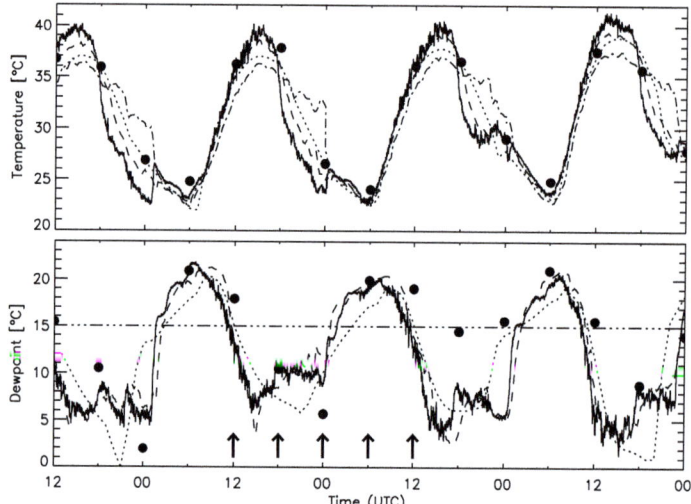

Figure 6.4: Time series of temperature (top) and dewpoint (bottom) close to the ground from 9 April 2006, 12 UTC to 13 April 2006, 00 UTC. Solid lines: HATPRO in-situ temperature and humidity sensor observations in Nangatchori (1.2 m AGL). Dashed lines: weather station observations in Djougou (2 m AGL). Dash-dotted line: lowest level of HATPRO retrieval for temperature (0–25 m AGL). Dotted lines: Méso-NH results for temperature and dewpoint at 10 m AGL. Black dots: ECMWF analyses. The 15 °C dewpoint line corresponds roughly to the position of the ITD and shows whether Nangatchori is north (drier) or south (moister) of the ITD. Arrows represent times of further comparisons in Section 6.5.

6 Detailed investigation of the ITD diurnal cycle over Nangatchori

The time series of dewpoint temperature demonstrate the large humidity contrasts within the diurnal cycle (Fig. 6.4, bottom). The 2 m dewpoint fluctuates between 5 °C and 20 °C which corresponds to a water vapor mixing ratio between 5 and 15 g kg^{-1}. The model reproduces well the main features of the diurnal cycle, such as the large contrasts of water vapor and temperature across the ITD (Fig. 6.4). Only the timing of the moist air arrival is delayed by one to two hours and it takes longer until the maximum dewpoint is reached, likely due to the fact that the model is too dry in the evening. The model does reproduce the jumps of humidity, but due to the lower temporal resolution (1 hour), they are less sharp. The diurnal variability of the atmosphere over Djougou and Nangatchori was better reproduced by Méso-NH than by the ECMWF analyses; e.g. difficulties in getting the minimum of dewpoint in the analyses on 11 April are displayed in Fig. 6.4.

The connection between dewpoint (Td) and ITD is given by the value of Td=15 °C (Buckle, 1996) which corresponds to the nighttime position of the ITD on surface charts. Examining the dewpoint diurnal cycle during the study period (Fig. 6.4, bottom), it can be seen that Nangatchori is equally influenced by both the Harmattan and the monsoon air mass in the course of the days. From about 00 to 12 UTC, the ITD is north of Djougou and moist air is present whereas dry air prevails from 12 to 00 UTC when the ITD is south of Djougou.

6.4.2 Ground-based remote sensing observations

Having shown the representativity of the ground-based values, we compare model time series of temperature, humidity and wind profiles with profiler observations from Nangatchori from 9 April 2006 12 UTC to 13 April 2006 00 UTC (Fig. 6.5). The diurnal cycle of temperature below 2000 m MSL from the microwave profiler measurements is well represented by the model. A particular characteristic of this diurnal (potential) temperature cycle (Fig. 6.5 a) is that the cooling in the range between 500 and 1000 m MSL starts only between 21 and 00 UTC in both the model and the observations. Sunset was at 1803 UTC which means that for the first 3 to 6 hours after sunset the radiative cooling does only affect a very thin layer (∼10 m) above the ground, in agreement with the previous section. At the time of the ITD arrival, the temperature drop in the measurements is very sharp, indicating the replacement of the warm Harmattan air mass by the cool monsoon flow around 00 UTC. The observations in layers above 2000 m MSL are slightly cooler than in the model and the well-mixed PBL is less deep in the observed temperature field. This might be due to the decreasing accuracy of microwave radiometer observations with height. As already mentioned before, the strong temperature change across the ITD is less sharp in the model than in the observations which can be seen very well when regarding the 310 K contour line (Fig. 6.5 a).

The abrupt change of air masses around midnight can also be well recognized in Fig. 6.5 b), where modelled and observed profiles of water vapor mixing ratio are presented. Note

6.4 Dynamics and thermodynamics of the ITD

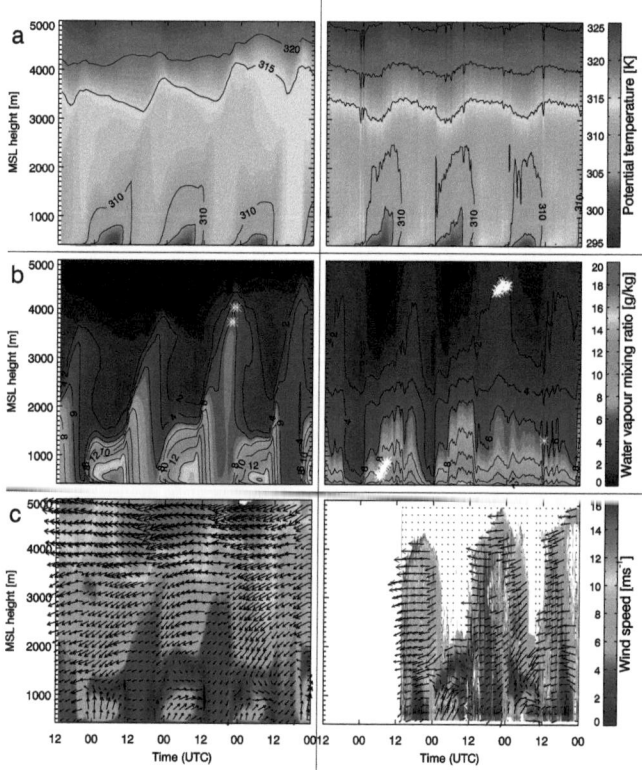

Figure 6.5: a): Time-height potential temperature cross sections over Nangatchori from 9 April 2006, 12 UTC to 13 April 2006, 00 UTC. Left: Méso-NH calculations. Right: HATPRO microwave profiler observations. b) Same as a), but for water vapor mixing ratio. White asterisks show 100 % relative humidity (calculated by model, left) and cloud base height detected by the ceilometer (right). c) Same as a), but for horizontal wind speed (left: Méso-NH results, right: UHF wind profiler). Arrows depict horizontal wind. Height information of UHF profiler is limited by the number of scattering particles. The time of ITD arrival was close to 00 UTC on all days.

6 Detailed investigation of the ITD diurnal cycle over Nangatchori

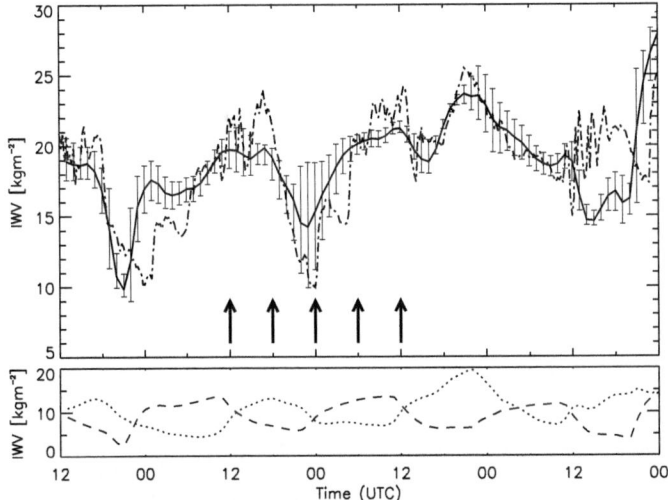

Figure 6.6: Time series of IWV over Nangatchori from 9 April 2006, 12 UTC to 13 April 2006, 00 UTC.
a) Solid line: Méso-NH output. Dash dotted line: HATPRO observations. Vertical bars indicate IWV values one grid box (i.e. 10km) further north (lower values) and south (higher values). Arrows represent times of further comparisons in Section 6.5.
b) The dashed line represents the modelled IWV within the lowest 1000 m AGL, the dotted line corresponds to the IWV above 1000 m AGL.

that the observations smear minima and maxima due to the poorer vertical resolution (see chapter 4). During the afternoon and until midnight, very dry air is advected by the Harmattan flow throughout the entire lower troposphere (up to 3500–4000 m MSL). This moisture decrease in the whole troposphere is also confirmed by the IWV time-series in Fig. 6.6. At midnight, these dry air masses are replaced by the moist air from the south within only a few minutes, resulting in quadrupled (from 4 to 16 g kg^{-1}) water vapor content close to the ground. During nighttime, the water vapor which has been transported into the region from the south is constrained to the lowest 1000 m AGL. After sunrise, the incoming solar radiation initiates vertical mixing of the atmosphere and in the course of the day, the water vapor becomes well mixed over the depth of the PBL which reaches up 3500 - 4000 m MSL. Due to the southward monsoon retreat, dry air advection sets in again.

6.4 Dynamics and thermodynamics of the ITD

On Fig. 6.5 b, the cloud base heights observed by the ceilometer reveal the presence of clouds on 10 April during the morning hours in the monsoon air layer close to the ground and also on 11 April around 20–23 UTC in 4500 m MSL on top of the residual boundary layer. For the later time, the model simulated a relative humidity of 100 % at the altitude where these mid-level clouds have been observed. These cloudy periods also correspond to the mixing ratio maximum from HATPRO profiles at those heights for the whole period of concern.

Fig. 6.5 c gives an overview of the wind conditions. The diurnal cycle can be seen very well with maximum wind speed in the PBL around sunrise which took place at 0545 UTC. The low-level wind maximum on 11 April between 00 and 06 UTC from the model is weaker ($10\,\mathrm{m\,s^{-1}}$) than the observed values by the wind profiler ($14\,\mathrm{m\,s^{-1}}$). This might not be a general model deficiency since on the last day of the study (13 April 00 UTC) the low-level jet seems to be well represented. In the afternoon and early evening, horizontal winds turn to north-easterlies within the whole low troposphere. Due to the higher surface temperature in the dry Harmattan air mass, the PBL reaches a depth of up to 4000 m and strong vertical momentum transport weakens horizontal winds significantly. This appears both in model simulations and observations.

The strong humidity changes in the whole atmosphere can be illustrated by considering the IWV (i.e. the total water vapor content in an atmospheric column) which varies between 10 and 28 $\mathrm{kg\,m^{-2}}$ during the period of concern (Fig 6.6). In order to present the impact of slight spatial shifts in the model, the closest grid box values to the north and south can be considered as error bars. Although the model grid spacing is only 10 km, differences of up to $9\,\mathrm{kg\,m^{-2}}$ over three grid boxes (30 km) can be seen, which confirms the presence of strong gradients over the area and the capability of the model to represent them. Most striking is the large diurnal variation. On 9 April, the observed IWV drops from 21 to $10\,\mathrm{kg\,m^{-2}}$ between 18 and 00 UTC. This means that within six hours the water vapor in the atmosphere was reduced by a factor of 2, which confirms the advection of very dry air masses in the early evening. The model calculations (solid line) also show a similar cycle, although the modelled IWV minimum occurred two hours earlier than observed. During the next night (10/11 April), the IWV dropped again from 23 to $10\,\mathrm{kg\,m^{-2}}$ (HATPRO observations). Note that the diurnal cycle of water IWV is not as marked as for near-surface variables due to the different vertical distribution of water vapor throughout the diurnal cycle. Model results reveal that from about 00 to 12 UTC roughly two thirds of the water vapor can be found below 1000 m AGL, whereas during the afternoon until midnight, two thirds of the water vapor are above 1000 m AGL (Fig. 6.6 b).

From the time series of wind, temperature and humidity profiles it is obvious that the diurnal boundary layer development is substantially influenced by the ITD diurnal cycle. The cooling after sunset is delayed due to warm and dry north-easterly winds, whereas after the ITD arrival the lowest 1000 m AGL are dominated by the moist and cooler monsoon air (e.g. Bou Karam et al., 2008). After sunrise, the convective boundary layer

6 Detailed investigation of the ITD diurnal cycle over Nangatchori

grows rapidly and the moisture is mixed into the whole PBL up to 4000 m AGL. In the afternoon north-easterly winds again bring very dry air masses along and the moist air is both mixed up into higher levels and transported back southwards.

6.4.3 Satellite observations

A common way to evaluate atmospheric models is by using satellite observations. These data have the great advantage of covering a more or less large area of the Earth at a glance. For example, Söhne et al. (2008) present a verification of Méso-NH cloud cover forecasts over West Africa during peak monsoon season. However, for boundary layer studies of temperature or humidity, the use of satellite data is usually very limited because the shape of the weighting function does not provide detailed profile information of the lower atmosphere. Furthermore, clouds in the thermal infrared and visible or water vapor at higher microwave channels obscure the view to the surface.

In this special case, though, MSG SEVIRI satellite observations turned out to be rather useful. When looking at nighttime infrared satellite images for cloudless nights in April 2006, they reveal a distinct north-south BT temperature gradient over the West African continent (Fig. 6.7). This feature which moves northward during the night has been recognized as the ITD position by comparison with ground-based observations. North of the ITD in the dry air, strong temperature inversions close to the ground (see Fig. 6.4) lead to low surface temperatures (i.e. cold IR BTs). In the example of 11 April 02 UTC (Fig. 6.7) many detailed structures show up north of about 10 deg. They reveal strong temperature inversions in basins and valleys, whereas hills and mountain ridges stayed above these shallow cold air pools even though the height difference is not more than 100 m. This is a well-known phenomenon on clear and dry nights with light winds (Gustavsson et al., 1998). South of the ITD, the radiation balance at the ground is different: because of the much higher water vapor content (and therefore stronger downwelling thermal radiation) as well as stronger winds, the surface remains warmer—even without clouds. This contrast in surface temperature across the ITD results in BT differences of about 3 K over only 10 km distance (Fig. 6.8). At 00 UTC, this BT jump lies exactly over Nangatchori (9.65 °N). From the model, unfortunately no surface skin temperature as analogue to satellite BT is available for direct comparison. However, we can use the lowest level (10 m) temperature to diagnose the ITD position. South of the ITD, satellite BT and 10 m temperature agree to about 4 K while north of 10 °N the BT is more than 8 K lower than the modelled 10 m temperature due to strong surface cooling (see Fig. 6.4). In order to derive the ITD position from the model, it turned out to be useful to look for the lowest 10 m temperature on the meridional section. Here the assumption is that the lowest temperature occurs at the convergence zone (lowest wind speed, see Fig. 6.3) and therefore strongest surface cooling. The ITD position in the model can then be derived at 9.9 °N at 00 UTC.

6.4 Dynamics and thermodynamics of the ITD

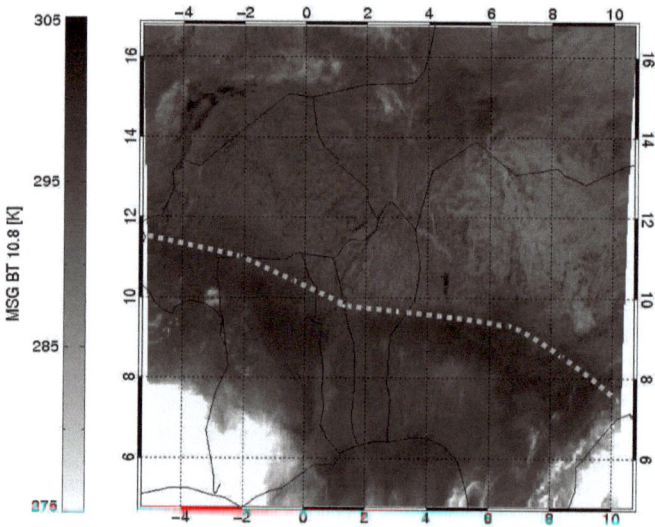

Figure 6.7: MSG SEVIRI 10.8 µm BT observations on 11 April 2006, 02 UTC. Dotted line corresponds to the approximate ITD position derived from Méso-NH results and ECMWF analyses by looking for the 15° dewpoint contour line.

To compare the speed of the ITD displacement from satellite and model, we looked first for time of the maximum meridional gradient of satellite BT every 0.1° from 8° to 11° N over a cross section between 1.5° and 2° E. In the model, the ITD position is diagnosed by looking for the latitudinal position where the strongest 10 m temperature decrease within one hour occurred. This is possible because the modelled 10 m temperature north of the ITD is largely influenced by the warm Harmattan flow and thus higher than south of the ITD (see also section 6.4.1). From Fig. 6.9 it is then possible to calculate the speed of the ITD displacement which is about 35 kilometers (0.3 degrees) per hour. The mean speed over the whole cross section (8 °N to 11 °N) is 9.98 m s^{-1} for MSG observations and 8.70 m s^{-1} for Méso-NH. When taking into account only the area within about 70 km from Nangatchori (9.25 °N to 10.25 °N), the observations (11.4 m s^{-1}) differ considerably from the model (8.2 m s^{-1}). This is also confirmed by the wind profiles (Fig. 6.5 c) where the UHF profiler observations show higher values than the model within the low-level jet. Compared to the observations, the ITD arrival in the model is about one hour too early between 8.5 °N and 9.5 °N, but model and satellite observations agree in the timing

6 Detailed investigation of the ITD diurnal cycle over Nangatchori

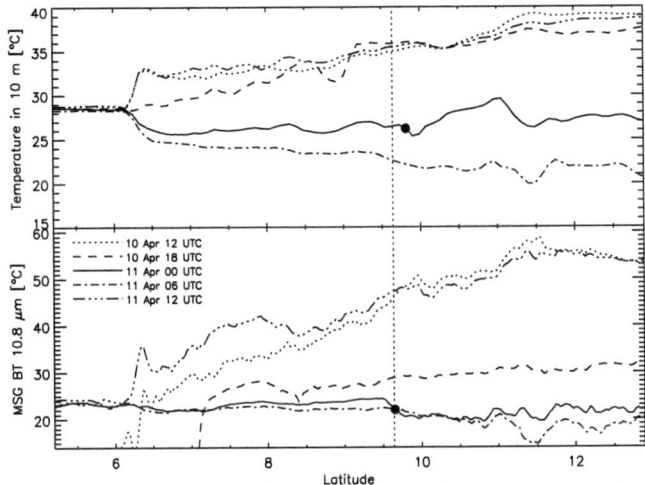

Figure 6.8: Top: East-west averaged temperature at 10 m AGL (first Méso-NH level) as a function of latitude, averaged over six grid boxes between 1.55 °E and 2.0 °E at different times on 10 and 11 April 2006. Bottom: Observed MSG infrared BTs of 10.8 µm channel (averaged over 1.5 °E and 2 °E). Vertical dashed line indicates the position of Nangatchori.

of ITD arrival north of 10 °N. The time of the BT jump over Nangatchori is also in line with the modelled wind jump (Fig. 6.5) giving a consistent timing (00 UTC) of the ITD passage. This confirms that the timing of the ITD arrival is simulated quite well in this night (10/11 April), except for the water vapor. In Section 6.4.1 we introduced the 15 °C dewpoint contour line as another criterion for the ITD position. As already mentioned, on 11 April this value is only reached about 3 hours after the observed ITD passage (Fig. 6.4) because the sharpness of the water vapor gradient is not captured properly and the establishment of the moist air takes more time in the model. The reason for that is not quite clear, but it could be related to the representativeness of the surface characteristics (humidity and soil type) in the model.

Although the surface skin temperature represented by the MSG observations is much higher during daytime compared to the Méso-NH 10 m temperature (Fig. 6.8), some distinct features agree very well, such as the gradual increase of both temperature and BT between 6 °N and 11 °N at 12 UTC. It has to be noted that on 10 April at 12 and 18 UTC some clouds were present south to 7 °N, resulting in lower BTs which are not represen-

81

6.5 A spatio-temporal view of the ITD diurnal cycle from model results

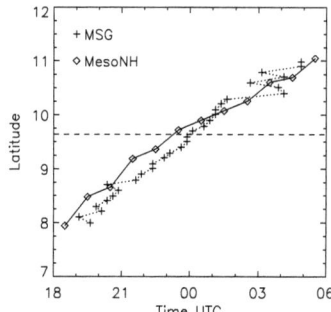

Figure 6.9: Solid line: ITD position as a function of time over the whole Méso-NH latitudinal cross section (average over 1.6 °E to 2.1 °E) on 10/11 April 2006. The position was diagnosed from Méso-NH output by looking for the latitudinal position of the strongest cooling in 10 m within one hour (solid line with diamonds). Data are plotted at xx30 UTC, assuming that the ITD displacement speed does not change much within one hour. Dotted line with crosses: Time of maximum MSG BT gradient over the latitudinal cross section between 8 °N and 11 °N (average over 1.6 °E and 2.1 °E). Horizontal dashed line indicates the position of Nangatchori.

tative of the surface conditions. Also interesting is the cooling between 00 and 06 UTC which increases with the latitude since the vegetation is sparser and it is much drier to the north. This enhances the nocturnal radiative cooling at night. Unfortunately, due to the frequent presence of clouds, the application of satellite data for routine detection of the ITD is limited. During the month of April 2006, only 3 nights were completely cloud-free over the area of Nangatchori (7, 9, and 10 April).

6.5 A spatio-temporal view of the ITD diurnal cycle from model results

Having found that the model is able to represent reality rather well, further analyses concerning the diurnal cycle can be made on the basis of the full 3-D information available by the model. First, the meridional variation of 925 hPa wind speed, pressure difference relative to Nangatchori and IWV for different times of the day are presented (Fig. 6.10). South of Nangatchori, low-level winds are generally weak during daytime. At sunset (18 UTC), the largest pressure difference relative to Nangatchori with 4 hPa can be seen between 6 °N and 9 °N. Although this pressure gradient is also visible at 12 UTC, strong

6 Detailed investigation of the ITD diurnal cycle over Nangatchori

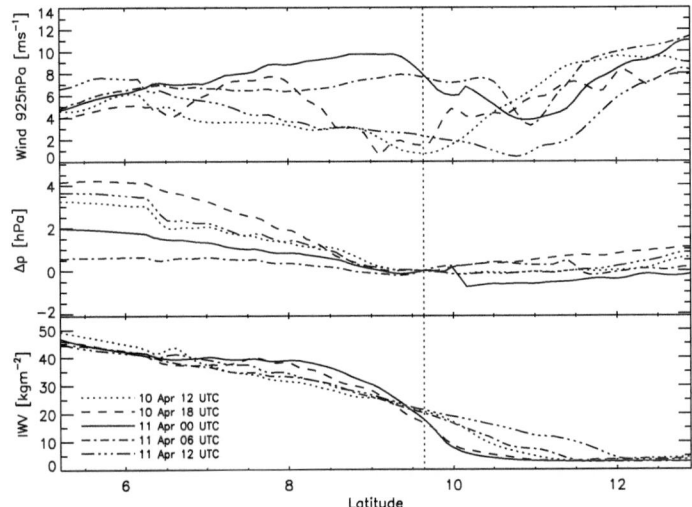

Figure 6.10: Méso-NH meridional cross sections, averaged over six grid boxes between 1.55 °E and 2.0 °E at different times on 10 and 11 April 2006. Vertical dashed line indicates the position of Nangatchori. Top: Wind speed in 925 hPa level. Center: Pressure difference relative to Nangatchori pressure value. Bottom: Integrated water vapor.

vertical turbulent mixing prevents the northward transport of moist air during the day. Once vertical mixing disappears around sunset, the downward momentum flux stops. This, combined with the loss of friction, enables the moist air in low levels to response to the pressure gradient force, producing a wind maximum between 7 °N and 8 °N at 18 UTC. The time around sunrise also corresponds to the time of the heat low maximum (Racz and Smith, 1999).

The strongest wind in the moist air mass is at 00 UTC on 11 April with its maximum at 9 °N, i.e. just south of the ITD position (the minimum of wind speed being directly over Nangatchori at that time). Some distance from the ITD (north of 12 °N) where the air remains dry for the whole period, 925 hPa winds are relatively constant both day and night (8 m s^{-1}). During the night the meridional surface pressure gradient weakens substantially and at 06 UTC it is less than 1 hPa over the whole area. After the ITD passage, the wind close to the surface declines slowly with the weakening pressure gradient due to friction (not shown). However, the (nearly frictionless) low-level jet in

6.5 A spatio-temporal view of the ITD diurnal cycle from model results

925 hPa remains strong until the morning hours and stops only when turbulent convective mixing sets in again after sunrise. The IWV (Fig. 6.10 c) shows its largest meridional gradient at 00 UTC and the moist air advancing further north later. The movement of the moisture gradient can also be identified well by diagnosing the water vapor mixing ratio contour line of $11\,\mathrm{g\,kg^{-1}}$ (Fig. 6.11), which corresponds to a dewpoint temperature of 15 °C at the surface (960 hPa). Between 10 April 18 UTC and 11 April 06 UTC, the $11\,\mathrm{g\,kg^{-1}}$ contour line (i.e. ITD position) moves about 200 km from 9 °N to 11 °N due to the nocturnal jet. During the same period, the vertical extent of the moist air mass diminishes. After one diurnal cycle the temperature and humidity conditions on 11 April 12 UTC are again very similar to those 24 hours before. Table 6.1 summarizes briefly the typical diurnal cycle for early April of atmospheric conditions at the Guinea Coast (6 °N), at Djougou (9.7 °N) and over the Sahel area (13–15 °N).

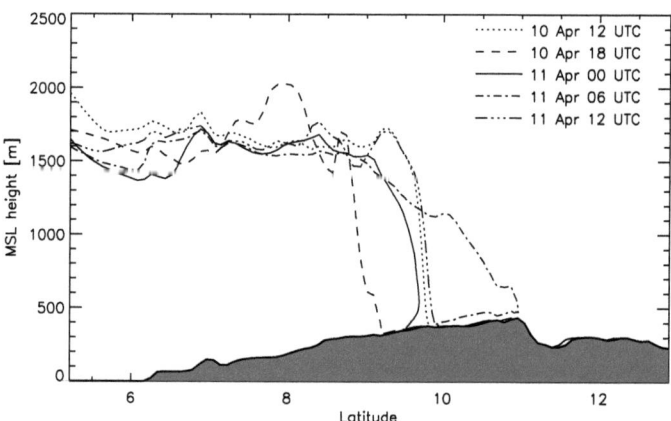

Figure 6.11: Meridional cross section of modelled $11\,\mathrm{g\,kg^{-1}}$ water vapor mixing ratio contour line at different times on 10 and 11 April 2006.

The modelled vertical distribution of water vapor at 00 UTC is illustrated by a meridional cross section (Fig. 6.12). Between the coast (6 °N) and Nangatchori (9.6 °N), the water vapor content close to the ground diminishes gradually. The strong gradient at 10 °N corresponds to the line of zero meridional wind speed. This line represents the border between the southerly monsoon flow (south of 10 °N and below 1300-1500 m MSL) and the northerly Harmattan flow (north of 10 °N and aloft of the monsoon flow).

6 Detailed investigation of the ITD diurnal cycle over Nangatchori

Table 6.1: Typical diurnal cycle over West Africa from the Gulf of Guinea to the Sahel during April 2006.

Time	Coast (6 °N)	Nangatchori (9.7 °N)	Sahel (13–15 °N)
12–18 UTC	Deep convection, large thundery cells with rain, moist air remains in coastal area	Dry convection, deep convective boundary layer (4 km), water vapor of last night is lifted from the ground to higher levels	Heat low with its center to the northwest of Benin (Mali, northern Ghana, Burkina Faso). Dry and hot northeasterly flow
18–00 UTC	"dying" convective cells, moist air starts to flow northwards	Dry Sahelian air in low levels. Boundary layer remains well-mixed except for the lowest < 100 m with a shallow temperature inversion	Pressure minimum remains over the area. At 00 UTC on 11 April it can be found over northwest Nigeria.
00–06 UTC	Southerly flow	Around midnight a quick jump from dry and warm air (easterly flow) to moist and relatively cool air (south-westerly flow). Stable stratification ($dT/dz \sim 0$ up to 800 m)	Pressure gradients weaken. Moist air penetrates into southern Sahel. Lowest pressure further north than at dawn.
06–12 UTC	Initiation of convection	Moisture supply from south continues, gradually onset of convection and rising convective boundary layer	Start of heat low formation. Dry air flow from north-east due to stronger pressure gradients

The model with its high vertical resolution allows investigating the development of the PBL using water vapor and temperature profiles. When regarding these pseudo-profiles (i.e. profiles from model simulations) over Nangatchori in Fig. 6.13, many of the characteristics of the West African monsoon can be identified. The modelled profiles provide a good overview over the diurnal cycle and the vertical structure of the PBL over Nangatchori during the time of ITD passage. The model results show very well the sharp contrasts between the dry and the moist side of the ITD, as well as between daytime (convective mixing) and nighttime (temperature inversions and low-level jet). The profile of potential temperature shows the extent of the PBL, which is 3000–3500 m deep at that time of the year. However, only at 18 UTC the whole layer is well-mixed. At 12 UTC, the mixed boundary layer reaches up only to 2000 m MSL and above this layer the residual layer can be seen. During nighttime, the atmosphere is stably stratified from

6.5 A spatio-temporal view of the ITD diurnal cycle from model results

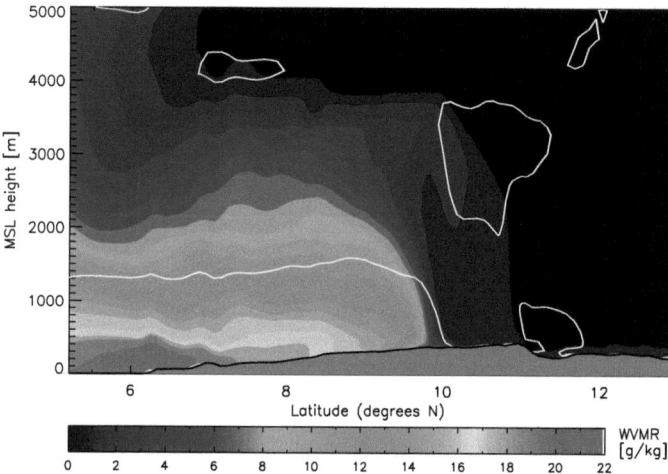

Figure 6.12: Meridional cross section of modelled water vapor mixing ratio at 1.8 °E on 11 April, 00 UTC. White contours represent zero meridional wind speed (which corresponds to the ITD position at the ground).

the ground to 2000 m MSL (at 00 and 06 UTC). Regarding the water vapor profiles, the well-mixed PBL at 18 UTC is represented by a constant mixing ratio value of $7 \, g \, kg^{-1}$. The strong inversion in 2000 m which was still present at 12 UTC is completely removed by (dry) convective mixing. At 00 UTC, the water vapor content in levels higher than 2000 m has been reduced due to horizontal advection. The highest mixing ratio over the whole cycle can be found at 06 UTC close to the ground ($15 \, g \, kg^{-1}$). This shallow moist layer corresponds to the south-westerly nocturnal jet which transports moisture to the north (Fig. 6.12). The highest values of water vapor transport occur at 06 UTC within the lowest 500 m AGL. Lothon et al. (2008) also identified the nocturnal jet maximum at about 05 UTC at about 450 m AGL using radiosonde and wind profiler observations from Niamey. The location of the jet peak around 450 m AGL explains that the zonal and meridional water vapor advection fluxes are maximum at this level. The presence of substantial wind shear favourites the mixing between the Harmattan and the monsoon layers and therefore contributes to the mixing of water vapor through the entire lower troposphere.

6 Detailed investigation of the ITD diurnal cycle over Nangatchori

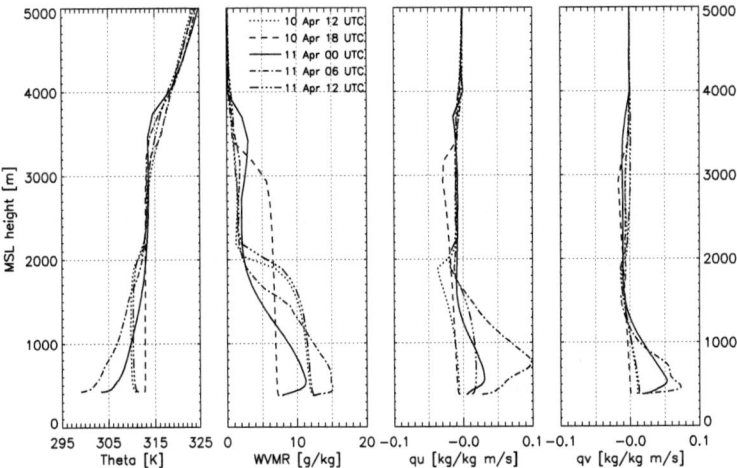

Figure 6.13: Model pseudo-soundings of potential temperature, water vapor mixing ratio, as well as the zonal (qu) and meridional (qv) water vapor advection fluxes over Nangatchori.

6.6 Statistical analysis of April 2006

Until here, this chapter focused on the three-day period for which the model was run. The nearly continuous ground-based observations in Nangatchori make it possible to perform a statistical analysis of the whole month of April 2006. During that month which encompasses the transition period from dry to wet conditions data availability of HATPRO temperature profiles is 83 % with 3169 boundary layer scans performed. This results in one profile every 11.3 minutes when averaging the number of scans over all available periods. Zenith observations providing IWV, LWP and humidity profiles are available for about 85 % of the month. For the ceilometer the data availability is even better with about 99 %. The reason for this difference in the data availability is the failure of the HATPRO instrument to restart after one specific power break which occurred on 22 April (see Fig. 6.14 a,c).

April 2006 was characterized by largely clear skies, i.e. only in 11.6 % of all measurements clouds below 7 km AGL were detected by the ceilometer (Fig. 6.14b). It has to be noted that clouds higher than this altitude or optical thin cirrus clouds (at altitudes above the freezing level at $\sim$4000 m) cannot be detected with this instrument. In the first period from 1–18 April, the cloud fraction was very low, which is typical for the dry season, only towards the end of the month the cloud amount increased, reaching up to 40 % on some days. Clouds below 3000 m (black bars in Fig. 6.14b) were mainly detected during the second part of the month.

The transition to a wetter climate is even more evident in IWV (Fig. 6.14d) whose mean value for the whole month is $34.6 \, \text{kg} \, \text{m}^{-2}$. Dry air was present in the beginning of the month which was replaced by quite humid conditions towards the end. The lowest 24-hour-mean occurred on 10 April with $18 \, \text{kg} \, \text{m}^{-2}$, whereas on 30 April a mean value of $50 \, \text{kg} \, \text{m}^{-2}$, i.e. much more than a doubling of IWV, could be observed.

Under the dominant clear sky conditions atmospheric stability is expected to show pronounced variations between very stable (nighttime) and unstable (daytime) conditions. Therefore the presence of temperature inversions was investigated using the following definition: For a given profile, at least one level between 50 and 1500 m AGL has to show a higher temperature than the lowest retrieved value at the ground. This criterion was met by 883 HATPRO temperature profiles (27.9 %). The maximum inversion strength during this period was 5.6 K/200 m which occurred on 7 April, 2330 UTC. (28.1 °C at the ground, 33.7 °C in 200 m AGL). This was shortly before the arrival of the moist air flow from the south which reached Nangatchori less than half an hour later. Generally, every night temperature inversions could be observed with highest values very often between 21 and 00 UTC. The absence of strong inversions in the early morning hours before sunrise where typically the strongest radiative cooling occurs can be explained by the following mechanism: The presence of dry air before midnight allows together with weak winds a more effective radiative cooling of the near surface air layer. Stronger winds

6 Detailed investigation of the ITD diurnal cycle over Nangatchori

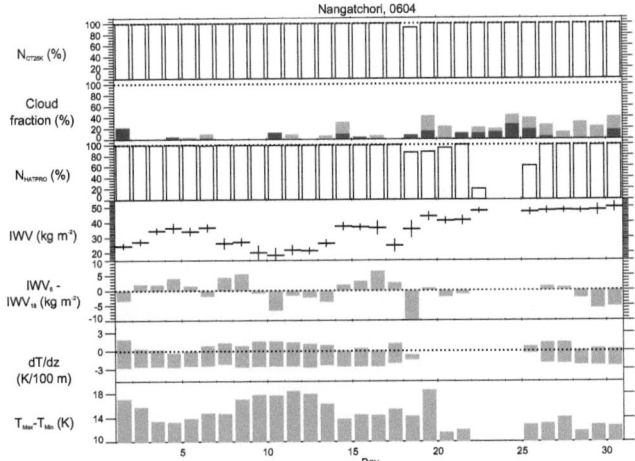

Figure 6.14: Statistical overview of Nangatchori measurements in April 2006. (a) Data availability of Ceilometer measurements per day (in %). (b) Fraction of the day with clouds detected by ceilometer (in %). In black lowest cloud base < 3 km AGL. Note: cloud bases higher than 7 km AGL are not detected (c) Data availability of HATPRO zenith observations per day (in %). (d) 24-hour mean of IWV plus standard deviation. (e) Difference between IWV at 06 UTC and IWV at 18 UTC. (f) Temperature difference between ground level and 200 m AGL. Bars show extremum values in 24-hour periods. Extremum values are mean of highest and lowest 10 % of measurements (g) Diurnal temperature range measured at the ground.

6.6 Statistical analysis of April 2006

and increased moisture prevent shallow temperature inversions to be formed after the passage of the nocturnal front.

With the nocturnal passage of the ITD, sharp drops in temperature associated with increasing atmospheric humidity were observed in many nights during the transition period. To quantify these jumps, a threshold of a 3 K temperature decrease within one hour at 200 m AGL was introduced. With this criterion, between 1 and 18 April, in 16 out of 18 nights this feature could be observed. The highest temperature changes (7.1 K within one hour) were detected in the nights 9/10 and 10/11 April, the mean value of temperature jumps over all 16 nights being 5 K. The mean front arrival was at 2319 UTC, the extremum values are 2034 UTC and 0220 UTC, respectively.

Examining more closely the Figs. 6.14d, e, f, and g, a five day period of moister conditions (3–7 April) is replaced by drier air masses (8–12 April). The daily mean water vapor content (Fig. 6.14d) correlates well with observed stability and diurnal temperature range. During the drier periods the vertical temperature gradient and the diurnal temperature range (Figs. 6.14f and g) is more pronounced. The time of arrival of the nocturnal front is also related to the mean daily IWV content. In the period 3–7 April, the mean front passage at Nangatchori was at 2200 UTC, whereas from 8–12 April the front arrived at 0020 UTC (not shown). One can assume that the ITD position in the evening was further south during the drier period, causing a later arrival of the front in Nangatchori. Fig. 6.14e gives another interesting result: In the moister period, at 18 UTC the IWV content was mostly higher than at 06 UTC whereas on the drier days, this diurnal variation is reversed. If daytime dry air inflow occurs IWV values at 18 UTC are lower than at 06 UTC. In contrast, if humid air is prevalent the IWV values tend to be higher in the evening due to daytime evaporation and possibly further advection of moist air from the south. It can be seen that the period of the case study presented earlier in this chapter (9–13 April) corresponds to the driest phase of the month. It should be noted that on moister days the north-easterly dry air inflow is not as strong as during the driest phase and towards the end of the month it is even completely missing due to the establishment of the monsoon flow.

7 Conclusions and Outlook

One main focus of the AMMA campaign was to enhance the network of atmospheric observations in West Africa, both by densifying the existing observation network and by applying new measurement methods, in order to better understand the monsoon system over West Africa. As part of this work, three remote sensing instruments (a microwave radiometer, a ceilometer and a micro rain radar) were operated for one year in Nangatchori (Benin) in the frame of the AMMA ground-based observation network. Despite the partly adverse conditions, such as frequent power failures, lack of internet connection, extensive heat or moisture, the instruments provided a unique data set covering one entire annual cycle of the West African climate with a data availability of more than 75 % of the time.

Especially within the low atmospheric layers, a new insight into parameters such as temperature, humidity and wind profiles was given, as satellites do not provide reliable information on the lower parts of the atmosphere. With the high temporal resolution, it was possible to get a continuous view of the processes. The comprehensive data set over a full year cycle enabled to investigate atmospheric processes via detailed case studies (chapter 6) but also on a statistical basis (section 6.6) Both approaches can be used to evaluate the performance of atmospheric models in this still data sparse region. While case studies might be more suitable to analyze whether mesoscale models are able to reproduce the small-scale variability observed, the statistical analysis enables a long-term evaluation of global models.

The analysis of the annual cycle of PBL properties exhibits well the two distinct seasons (dry and wet) with their sharp contrasts. However, the "dry" season in early 2006 was characterized by several moist air outbreaks from the south, resulting in relatively moist and warm average conditions. In contrast, January 2007 brought much cooler and drier conditions due to a strong Harmattan season. Also the wet season shows large inter-annual variability, especially when regarding the rainfall patterns over the northern parts of the study region. Therefore, for extended statistical analyses and deeper knowledge of inter-annual variations of the ITD position and its diurnal cycle, long-term PBL observations as well as further model studies are necessary.

A detailed investigation of the ITD characteristics shows the potential of the mesoscale atmospheric model Méso-NH which was run for a 84-hour period in April 2006 to describe mesoscale lower tropospheric features like low-level jets associated with the diurnal cycle of the ITD. It was a challenge for the model to reproduce the sharp contrasts across the ITD properly because of the low number of observations (esp. radiosondes) from this data sparse region that were assimilated in the model analysis, but thanks to the availability of high-quality observational data from Nangatchori, it is possible to validate the model. An evaluation of the model results shows that in particular the strong temperature and humidity gradients between the moist monsoon air and the dry Harmattan air turned out to be well simulated. The water vapor mixing ratio close to the ground fluctuated between 5 and 15 g kg^{-1} within a diurnal cycle. IWV changes from 10 to 28 kg m^{-2}. A comparison with radiosonde ascents from Parakou ($\sim$100 km southeast of Nangatchori) also confirms the good performance of the model during daytime. Unfortunately, radiosondes were only launched at 12 UTC during April 2006 and therefore no information on the diurnal cycle is available from these data.

As a novel approach, the modelled 2 m air temperature has been compared to MSG infrared BT observations at 10.8 μm for cloud-free nights around the ITD position. In the dry air north of the ITD, a strong shallow nighttime temperature inversion caused lower surface temperatures (and thus lower satellite BT) than south of the ITD where a much higher IWV resulted in reduced nighttime cooling. From the northward movement of the strongest BT gradient (which is located at the ITD), it was possible to obtain the mean nighttime ITD displacement speed between 8 °N and 11 °N (10.0 m s^{-1}) from the satellite data. Despite the restriction through clouds, the potential of satellites for detecting low-level atmospheric features over this area with only few ground-based and upper-air observations should be studied further.

The ground-based remote sensing instrumentation at the supersites of Nangatchori and Niamey covered many interesting atmospheric features. However, e.g. radiosondes directly at the Nangatchori site or a cloud radar would have been ideal complementary tools to the instrument setup. This setup would then allow deriving physically consistent temperature, humidity and cloud liquid water profiles of the atmosphere (Löhnert et al., 2007).

Many of the remote sensing instruments deployed during AMMA were state-of-the-art products and the development of new features is an ongoing process for these devices. For example, during the last two years, some HATPRO microwave radiometers have been equipped with an additional azimuth scanning ability. The instrument at Nangatchori was only able to perform elevation scans to one fixed azimuth direction. The scans have been performed every 10–15 minutes in northward direction. The azimuth scanning now enables 3D scans which have until now mainly used to detect spatial water vapor variations (Kneifel et al., 2009). A glimpse on the potential for detecting nocturnal fronts at Nangatchori is given in Fig. 7.1 where the BTs along the water vapor line under two different elevation angles are presented. The moist air arrives earlier when observing under an elevation angle of 70° and only about 15 minutes later when looking with a low elevation angle of 5.4° to the north. If the azimuth scanning possibility had already been available at Nangatchori, the arrival of the moist air could likely have been documented

7 Conclusions and Outlook

in even more detail.

Hundreds of scientists of different fields were directly or indirectly involved in the AMMA project and it was a huge effort to capture the characteristics of the West African monsoon system. Overall, the observations in Africa were very successful and until now already a large number of exciting research articles has appeared from the completely new results gained during the field campaigns. This work shall contribute to the atmospheric research within the AMMA project and will hopefully lead to a better understanding of the West African monsoon and its role within the global climate system.

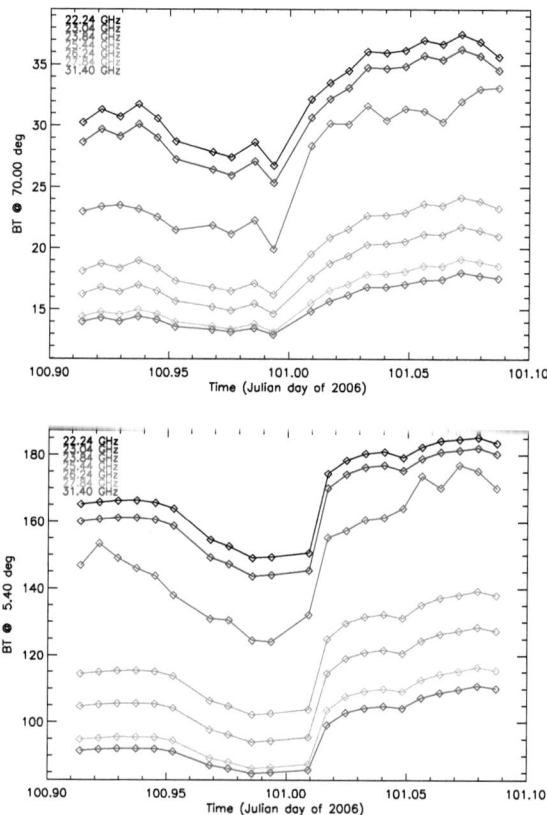

Figure 7.1: 10/11 April 2006, ITD arrival. Top: BTs along the water vapor line at 70° elevation angle. Bottom: Same for 5.4° elevation angle (pointing to north).

Bibliography

Ali, A. and T. Lebel (2009). The Sahelian standardized rainfall index revisited. *Int. J. Climatol. 29*, 1705–1714.

AMMA-EU (2007). D6.1.e: Report on data submitted to database. Technical report, AMMA-EU.

Atlas, D., R. C. Srivastava, and R. S. Sekhon (1973). Doppler Radar Characteristics of Precipitation at Vertical Incidence. *Rev. Geophys. Space Phys. 11*, 1–35.

Bechtold, P., E. Bazile, F. Guichard, P. Mascart, and E. Richard (2001). A mass-flux convection scheme for regional and global models. *Q. J. Roy. Meteor. Soc. 127*, 869–886.

Blackadar, A. K. (1957). Boundary layer wind maxima and their significance for the growth of nocturnal inversions. *B. Am. Meteor. Soc. 38*, 283–290.

Bock, O., M. N. Bouin, E. Doerflinger, P. Collard, F. Masson, R. Meynadier, S. Nahmani, M. Koité, K. Gaptia Lawan Balawan, F. Didé, D. Ouedraogo, S. Pokperlaar, J. B. Ngamini, J. P. Lafore, S. Janicot, F. Guichard, and M. Nuret (2008). The West African Monsoon observed with ground-based GPS receivers during AMMA. *J. Geophys. Res. 113(D21105)*, doi:10.1029/2008JD010327.

Bou Karam, D., C. Flamant, P. Knippertz, O. Reitebuch, M. Chong, J. Pelon, and A. Dabas (2008). Dust emissions over the Sahel associated with the West African Monsoon inter-tropical discontinuity region: a representative case study. *Q. J. Roy. Meteor. Soc. 134*, 621–634.

Bougeault, P. and P. Lacarrère (1989). Parametrization of orography-induced turbulence in a meso-beta model. *Mon. Wea. Rev. 117*, 1872–1890.

Buckle, C. (1996). *Weather and climate in Africa*. Addison-Wesley Longman Ltd, Harlow, UK.

Couvreux, F., F. Guichard, O. Bock, B. Campistron, J. P. Lafore, and J. L. Redelsperger

Bibliography

(2009). Synoptic variability of the monsoon flux over West Africa prior to the onset. *Q. J. Roy. Meteor. Soc. in press*, doi:10.1002/qj.473.

Crewell, S. and U. Löhnert (2003). Accuracy of cloud liquid water path from ground-based microwave radiometry. Part 2: Sensor accuracy and synergy. *Radio Sci. 38*, 8041, doi:10.1029/2002RS002634.

Crewell, S. and U. Löhnert (2007). Accuracy of Boundary Layer Temperature Profiles Retrieved With Multifrequency Multiangle Microwave Radiometry. *IEEE T. Geosci. Remote. 45*, 2195–2201.

Dolman, A. J., A. D. Culf, and P. Bessemoulin (1997). Observations of boundary layer development during the HAPEX-Sahel intensive observation period. *J. Hydrol. 189*, 998–1016.

Drobinski, P., S. Bastin, S. Janicot, O. Bock, A. Dabas, P. Delville, O. Reitebuch, and B. Sultan (2009). On the late northward propagation of the West African monsoon in summer 2006 in the region of Niger/Mali. *J. Geophys. Res. 114*, doi:10.1029/2008JD011159.

Fink, A., D. G. Vincent, and V. Ermert (2006). Rainfall Types in the West African Sudanian Zone during the Summer Monsoon 2002. *Mon. Wea. Rev. 134*, 2143–2164.

Flamant, C., P. Knippertz, D. J. Parker, J. P. Chaboureau, C. Lavaysse, A. Agusti-Panareda, and L. Kergoat (2009). The impact of a mesoscale convective system cold-pool on the northward propagation of the inter-tropical discontinuity over West Africa. *Q. J. Roy. Meteor. Soc. 135*, 139–159.

Garratt, J. R. (1994). *The Atmospheric Boundary Layer*. Cambridge University Press.

Güldner, J. and D. Spänkuch (2001). Remote Sensing of the Thermodynamic State of the Atmospheric Boundary Layer by Ground-Based Microwave Radiometry. *J. Atmos. Ocean. Tech. 18*, 925–933.

Gustavsson, T., M. Karlsson, J. Bogren, and S. Lindqvist (1998). Development of Temperature Patterns during Clear Nights. *J. Appl. Meteor. 37*, 559–571.

Hastenrath, S. (1985). *Climate and Circulation of the Tropics*. D. Reidel Publishing Company.

Heo, B. H., S. Jacoby-Coaly, K. E. Kim, B. Campistron, B. Bénech, and E. S. Jung (2003). Use of the Doppler spectral width to improve the estimation of the convective boundary layer height. *J. Atmos. Ocean. Tech. 20*, 408–424.

Bibliography

Jacoby-Coaly, S., B. Campistron, S. Bernard, B. Bénech, F. Ardhuin-Girard, J. Dessens, E. Dupont, and B. Carissimo (2002). Turbulent dissipation rate in the boundary layer via UHF wind profiler doppler spectral width measurements. *Boundary-Lay. Meteorol. 103*, 361–389.

Janicot, S., C. D. Thorncroft, A. Ali, N. Asencio, G. Berry, O. Bock, B. Bourles, G. Caniaux, F. Chauvin, A. Deme, L. Kergoat, J. P. Lafore, C. Lavaysse, T. Lebel, B. Marticorena, F. Mounier, P. Nedelec, J. L. Redelsperger, F. Ravegnani, C. E. Reeves, R. Roca, P. de Rosnay, H. Schlager, B. Sultan, M. Tomasini, A. Ulanovsky, and A. forecasters team (2008). Large-scale overview of the summer monsoon over West Africa during the AMMA field experiment in 2006. *Ann. Geophys. 26*, 2569–2595.

Janssen, M. A. (1993). *Atmospheric remote sensing by microwave radiometry.* John Wiley & Sons, Inc.

Kain, J. S. and J. M. Fritsch (1993). *Convective parameterization for mesoscale models: The Kain-Fritsch scheme. The Representation of Cumulus Convection in Numerical Models.* American Meteorological Society, Boston.

Karstens, U., C. Simmer, and E. Ruprecht (1994). Remote Sensing of Cloud Liquid Water. *Meteorol. Atmos. Phys. 54*, 157–171.

Klett, J. D. (1981). Stable analytical inversion solution for processing lidar returns. *Appl. Opt. 20*, 211–220.

Kneifel, S., S. Crewell, U. Löhnert, and J. Schween (2009). Investigating Water Vapor Variability by Ground-Based Microwave Radiometry: Evaluation Using Airborne Observations. *IEEE Geosci. Remote Sens. Lett. 6*, 157–161.

Kottek, M., J. Grieser, C. Beck, B. Rudolf, and F. Rubel (2006). World Map of Köppen-Geiger Climate Classification updated. *Meteorol. Z. 15*, 1–5.

Köppen, W. and R. Geiger (1928). Klimakarte der Erde. Verlag Justus Perthes, Gotha.

Lafore, J. P., J. Stein, N. Asencio, P. Bougeault, V. Ducrocq, J. Duron, C. Fischer, P. Hreil, P. Mascart, V. Masson, J. Pinty, J. L. Redelsperger, E. Richard, and J. Vil-Guerau de Arellano (1998). The Méso-NH Atmospheric Simulation System. Part I: adiabatic formulation and control simulations. Scientific objectives and experimental design. *Ann. Geophys. 16*, 90–109.

Le Barbé, L., T. Lebel, and D. Tapsoba (2002). Rainfall Variability in West Africa during the years 1950-90. *J. Climate 15*, 187–202.

Lebel, T. and A. Ali (2009). Recent trends in the Central and Western Sahel rainfall regime (1990-2007). *Journal of Hydrology 375*, 52–64.

Lebel, T., D. J. Parker, B. Bourles, C. Flamant, B. Marticorena, C. Peugeot, A. Gaye, J. Haywood, E. Mougin, J. Polcher, J. L. Redelsperger, and C. D. Thorncroft (2009). The AMMA field campaigns: Multiscale and multidisciplinary observations in the West African region. *Q. J. Roy. Meteor. Soc. in press*, doi: 10.1002/qj.486.

Lebel, T., J. D. Taupin, and N. D'Amato (1997). Rainfall monitoring during HAPEX-Sahel. 1. General rainfall conditions and climatology. *Journal of Hydrology 188-189*, 74 – 96. HAPEX-Sahel.

Löhnert, U. and S. Crewell (2003). Accuracy of cloud liquid water path from ground-based microwave radiometry. Part I: Dependency on cloud model statistics and precipitation. *Radio Sci. 38*, 8041, doi:10.1029/2002RS002654.

Löhnert, U., D. D. Turner, and S. Crewell (2009). Ground-based temperature and humidity profiling using spectral infrared and microwave observations. Part 1. Retrieval performance in clear sky conditions. *J. Appl. Meteorol. Clim. 48*, 1017–1032.

Löhnert, U., E. van Meijgaard, H. K. Baltink, S. Groß, and R. Boers (2007). Accuracy assessment of an integrated profiling technique for operationally deriving profiles of temperature, humidity, and cloud liquid water. *J. Geophys. Res. 112*, doi:10.1029/2006JD007379.

Liljegren, J. C., S. A. Boukabara, K. Cady-Pereira, and S. Clough (2005). The Effect of the Half-Width of the 22-GHz Water Vapor Line on Retrievals of Temperature and Water vapor Profiles with a Twelve-Channel Microwave Radiometer. *IEEE Trans. Geosci. Reomote Sens. 43*, 1102–1108.

Lothon, M., F. Saïd, F. Lohou, and B. Campistron (2008). Observation of the diurnal cycle in the low troposphere over West Africa. *Mon. Wea. Rev. 136*, 3477–3500.

Mallet, M., V. Pont, C. Liousse, L. Gomes, J. Pelon, S. Osborne, J. Haywood, J. C. Roger, P. Dubuisson, A. Mariscal, V. Thouret, and P. Goloub (2008). Aerosol direct radiative forcing over Djougou (northern Benin) during the African Monsoon Multidisciplinary Analysis dry season experiment (Special Observation Period-0). *J. Geophys. Res. 113*, doi:10.129/2007JD009419.

Masson, V., J. L. Champeaux, F. Chauvin, C. Meriguet, and R. Lacaze (2003). A Global Database of Land Surface Parameters at 1-km Resolution in Meteorological and Climate Models. *J. Climate 16*, 1261–1282.

Miller, M. A. and A. Slingo (2007). The ARM Mobile Facility and Its First International

Bibliography

Deployment: Measuring Radiative Flux Divergence in West Africa. *B. Am. Meteorol. Soc. 88*, 1229–1244.

Moumouni, S., M. Gosset, and E. Houngninou (2008). Main features of rain drop size distributions observed in Benin, West Africa, with optical disdrometers. *Geophys. Res. Lett. 35*, doi:10.1029/2008GL035755.

Noilhan, J. and J. F. Mahfouf (1996). The ISBA land surface parameterization scheme. *Global and Plan. Change 13*, 145–159.

Parker, D. J., R. R. Burton, A. Diongue-Niang, R. J. Ellis, M. Felton, C. M. Taylor, C. D. Thorncroft, P. Bessemoulin, and A. M. Tompkins (2005). The diurnal cycle of the West African monsoon circulation. *Q. J. Roy. Meteor. Soc. 131*, 2839–2860.

Parker, D. J., A. Fink, S. Janicot, J.-B. Ngamini, M. Douglas, E. Afiesimama, A. Agusti-Panareda, A. Beljaars, F. Didé, A. Diedhiou, T. Lebel, J. Polcher, J.-L. Redelsperger, C. Thorncroft, and W. G. A. (2008). The AMMA radiosonde program and its implications for the future of atmospheric monitoring over Africa. *B. Am. Meteorol. Soc. 89*, 1015–1027.

Pelon, J., M. Mallet, A. Mariscal, P. Goloub, D. Tanré, D. Bou Karam, C. Flamant, J. Haywood, B. Pospichal, and S. Victori (2008). Microlidar observations of biomass burning aerosol over Djougou (Benin) during African Monsoon Multidisciplinary Analysis Special Observation Period 0: Dust and Biomass-Burning Experiment. *J. Geophys. Res. 113*, D00C18, doi:10.1029/2008JD009976.

Peters, G., B. Fischer, and T. Andersson (2002). Rain observations with a vertically looking Micro Rain Radar (MRR). *Boreal Env. Res. 7*, 353–362.

Pospichal, B. and S. Crewell (2007a). Boundary layer observations in West Africa using a novel microwave radiometer. *Meteorol. Z. 16*, 513–523.

Pospichal, B. and S. Crewell (2007b). Deliverable D2.1.A.a: Report on statistical description of the diurnal and annual cycle of various parameters (humidity, clouds, stability) in the Planetary Boundary Layer (PBL). AMMA Deliverable.

Pospichal, B. and S. Crewell (2007c). Report on AMMA SOP instrument deployment in Nangatchori/Djougou (Benin). Internal report, University of Cologne.

Racz, Z. and R. K. Smith (1999). The dynamics of heat lows. *Q. J. Roy. Meteor. Soc. 125*, 225–252.

Redelsperger, J. L., C. D. Thorncroft, A. Diedhiou, T. Lebel, D. J. Parker, and J. Polcher

Bibliography

(2006). African Monsoon Multidisciplinary Analysis: An International Research Project and Field Campaign. *B. Am. Meteorol. Soc. 87*, 1739–1746.

Rogers, R. R., M.-F. Lamoureux, L. R. Bissonnette, and R. M. Peters (1997). Quantitative Interpretation of Laser Ceilometer Intensity Profiles. *J. Atmos. Ocean. Tech., 14*, 396–411.

Rose, T., S. Crewell, U. Löhnert, and C. Simmer (2005). A network suitable microwave radiometer for operational monitoring of the cloudy atmosphere. *Atmos. Res. 75*, 183–200.

Saxton, J., A. Lewis, J. Kettlewell, M. Ozel, F. Gogus, T. Boni, S. Korogone, and D. Serça (2007). Isoprene and monoterpene measurements in a secondary forest in northern Benin. *Atm. Chem. Phys. 7*, 4095–4106.

Schmetz, J., P. Pili, S. Tjemkes, D. Just, J. Kerkmann, S. Rota, and A. Ratier (2002). An introduction to Meteosat Second Generation (MSG). *Bull. Amer. Meteor. Soc. 83*, 977–992.

Schrage, J., S. Augustyn, and A. H. Fink (2007). Nocturnal stratiform cloudiness during the West African monsoon. *Meteorol. Atmos. Phys. 95*, 73–86.

Söhne, N., J. P. Chaboureau, and F. Guichard (2008). Verification of cloud cover forecast with satellite observation over West Africa. *Mon. Wea. Rev. 136*, 4421–4434.

Sultan, B. and S. Janicot (2000). Abrupt shift of the ITCZ over West Africa and intraseasonal variability. *Geophys. Res. Lett. 27*, 3353–3356.

Sultan, B. and S. Janicot (2003). The West African Monsoon Dynamics. Part II: The "Preonset" and "Onset" of the Summer Monsoon. *J. Climate 16*, 3407–3427.

Sultan, B., S. Janicot, and P. Drobinski (2007). Characterization of the Diurnal Cycle of the West African Monsoon around the Monsoon Onset. *J. Climate 20*, 4014–4032.

Thorncroft, C., D. J. Parker, R. R. Burton, M. Diop, J. H. Ayers, H. Barjat, S. Devereau, A. Diongue, R. Dumelow, D. R. Kindred, N. M. Price, M. Saloum, C. M. Taylor, and A. M. Tompkins (2003). The JET2000 Project: Aircraft Observations of the African Easterly Jet and African Easterly Waves. *Bull. Amer. Meteor. Soc. 84*, 337–351.

Tulet, P., M. Mallet, V. Pont, J. Pelon, and A. Boone (2008). The 7 - 13 March 2006 dust storm over West Africa: Generation, transport, and vertical stratification. *J. Geophys. Res. 113*, doi:10.1029/2008JD009871.

Turner, D. D. (2007). Improved ground-based liquid water path retrievals using

Bibliography

a combined infrared and microwave approach. *J. Geophys. Res. 112, D15204*, doi:10.1029/2007JD008530.

Ware, R., R. Carpenter, J. Güldner, J. Liljegren, T. Nehrkorn, F. Solheim, and F. Vandenberghe (2003). A multichannel radiometric profiler of temperature, humidity, and cloud liquid. *Radio Sci. 38*, doi:10.1029/2002RS002856.

Westwater, E. R., Y. Han, V. G. Irisov, V. Leuskiy, E. N. Kadygrov, and S. A. Viazankin (1999). Remote Sensing of Boundary Layer Temperature Profiles by a Scanning 5-mm Microwave Radiometer and RASS: Comparison Experiments. *J. Atmos. Ocean. Tech. 16*, 805–818.

List of Figures

2.1 Climate zones of Africa (Köppen-Geiger classification) after Kottek et al. (2006). 11
2.2 Standardized precipitation index for the Sahel area. Source: Ali and Lebel (2009). 12
2.3 Meridional section across West Africa in August showing the position of the main airstreams, the monsoon trough, and the principal weather zones with the ITCZ at 15 °N and the ITD at 21 °N. Source: Buckle (1996). . . 13
2.4 Solid line: 10-day average ITD position along the 0 ° meridian (mean over the years 1988–2005). Dashed lines indicate the maximum and minimum position during this period. Source: NOAA Climate Prediction Center, African ITCZ Monitoring Project (http://www.cpc.noaa.gov/products/fews/ITCZ/itcz.shtml). 14

3.1 Space and time scales in AMMA. Source: Redelsperger et al. (2006). . . . 18
3.2 Nangatchori site in January 2006. Left: Tower with turbulence and flux measurements. Center: Lidar Ceilometer. Right: HATPRO microwave radiometer. 22
3.3 ARM mobile facility (AMF) in Niamey in November 2005 (shortly after deployment). In the front is the aerosol stack and the containers, in the background the AMF instrument field. Courtesy: U.S. Department of Energy's Atmospheric Radiation Measurement Program. 25

4.1 Atmospheric extinction in the microwave spectrum for a cloudy atmosphere. The red line shows the oxygen contribution, the green line the water vapor and the blue line the contribution of liquid water assuming a liquid water content (LWC) of $0.2\,\mathrm{g\,m^{-3}}$. The black line represents the sum of all components. Measurements by the HATPRO microwave radiometer are performed in the frequency bands A and B. 29
4.2 HATPRO BT on 11 June 2006 in Nangatchori. 31
4.3 24-hour timeseries of IWV and LWP from HATPRO observations in Nangatchori as well as the lowest cloud base (LCB) detected by the ceilometer on 11 June 2006. 32
4.4 Flow chart for the development of statistical retrieval algorithms for ground-based microwave radiometers from a radiosonde data set using a RT model and mulitilinear regression. 33
4.5 Ceilometer backscatter profiles over Nangatchori on 11 June 2006. 36

4.6 Radar reflectivity on 9 September 2006 as a function of height from MRR measurements. 38
4.7 Same as 4.6, but for 28 July 2006. 38
4.8 Same as 4.6, but for 23 March 2006. 39
4.9 1-hour mean IWV GPS vs. HATPRO microwave radiometer (whole year 2006). 40
4.10 HATPRO data availability in 2006. 42

5.1 IWV annual cycle (black) and daily rain accumulations (blue bars) in 2006. IWV daily mean values derived from microwave radiometer measurements. Data gaps are filled with GPS observations. a) Nangatchori/Djougou. b) Niamey. 44
5.2 Cumulative frequency distribution of IWV from HATPRO measurements at Nangatchori. Dry season: January–April and November–December 2006. Wet season: May–October 2006. 46
5.3 Daily percentage of cloudy scenes from ceilometer observations: a) Nangatchori/Djougou, b) Niamey. Black: percentage of day with clouds detected by ceilometer, grey: percentage of day without clouds, white: no data. 48
5.4 Annual cycle of daily maximum (T_{max}) and minimum (T_{min}) temperature measured in 2 meters above ground at Djougou and Niamey in 2006 (top) as well as diurnal temperature range (T_{max} minus T_{min}, bottom). 50
5.5 Diurnal 2m-temperature range (T_{max} minus T_{min}) as a function of daily cloud cover for all 2006. a) Nangatchori/Djougou. b) Niamey. Colors indicate the diurnal mean IWV value. 52
5.6 Diurnal 2m-temperature range (T_{max} minus T_{min}) as a function of daily mean long-wave radiation balance in Niamey for the whole year 2006. Colors indicate the diurnal mean IWV value. 53
5.7 Diurnal 2m-temperature range (T_{max} minus T_{min}) as a function of short-wave radiation balance (averaged over daytime between sunrise and sunset) in Niamey for the whole year 2006. Colors indicate the diurnal mean long-wave radiation balance. 53
5.8 Daily mean long-wave radiation balance as a function of daily mean IWV in Niamey for the whole year 2006. Colors indicate the diurnal temperature range. 54
5.9 Annual cycle of daily mean long-wave radiation balance, diurnal temperature range and daily mean IWV for Niamey in 2006. 55
5.10 Monthly mean diurnal cycle of potential temperature in 50 m AGL over Nangatchori from microwave radiometer observations. 56
5.11 Cloud base height frequency distributions from ceilometer observations for two months' periods at a) Nangatchori and b) Niamey. Top left: Jan/Feb, top right: Mar/Apr, center left: May/Jun, center right: Jul/Aug, bottom left: Sep/Oct, bottom right: Nov/Dec. 58

5.12 Diurnal cycle of rainfall for the period between 1 June and 15 September 2006. Black line represents the number of days where rainfall was observed during a certain hour. Blue bars show cumulated rainfall over the whole period. Top: Nangatchori, Bottom: Niamey. 59

5.13 Mean diurnal cycle of hourly mean IWV observed in Nangatchori/Djougou for different periods in 2006. Red: HATPRO, black: GPS. Dashed line includes also GPS measurements from times where no HATPRO measurements are available. 60

5.14 Mean diurnal cycle of LWP from HATPRO observations in Nangatchori (June–September 2006). Only cloudy times detected by the ceilometer were considered. 61

5.15 LWP versus cloud base height for 2006: a) Nangatchori, b) Niamey. 62

5.16 Diurnal cycle of potential temperature gradient between 700 and 50 m AGL from HATPRO observations for different months in 2006 and January 2007 in Nangatchori. 63

5.17 Mean monthly profiles of potential temperature. Top: 06 UTC, Center: 15 UTC, Bottom: 21 UTC. 64

5.18 Diurnal cycle of time-height cross sections of temperature, relative humidity, potential temperature and equivalent potential temperature for a) 12/13 April 2006 and b) 10/11 June 2006. Note that the range of all values changes between the events. 66

5.19 Same as Fig. 5.18, but for a) 8/9 August 2006 and b) 14/15 December 2006. 67

6.1 Méso-NH Model domain with orography (shaded). Djougou is situated 10 km northwest of Nangatchori. 70

6.2 Synoptic situation on 11 April 2006, 00 UTC from ECMWF analysis . . . 71

6.3 ECMWF analysis of 925 hPa wind speed on 11 April 2006, 00 UTC 72

6.4 Time series of temperature (top) and dewpoint (bottom) close to the ground from 9 April 2006, 12 UTC to 13 April 2006, 00 UTC. 74

6.5 a): Time-height potential temperature cross sections over Nangatchori from 9 April 2006, 12 UTC to 13 April 2006, 00 UTC. Left: Méso-NH calculations. Right: HATPRO microwave profiler observations. b) Same as a), but for water vapor mixing ratio. White asterisks show 100 % relative humidity (calculated by model, left) and cloud base height detected by the ceilometer (right). c) Same as a), but for horizontal wind speed (left: Méso-NH results, right: UHF wind profiler). Arrows depict horizontal wind. Height information of UHF profiler is limited by the number of scattering particles. The time of ITD arrival was close to 00 UTC on all days. 76

6.6 Time series of IWV over Nangatchori from 9 April 2006, 12 UTC to 13 April 2006, 00 UTC. a) Solid line: Méso-NH output. Dash dotted line: HATPRO observations. Vertical bars indicate IWV values one grid box (i.e. 10km) further north (lower values) and south (higher values). Arrows represent times of further comparisons in Section 6.5. b) The dashed line represents the modelled IWV within the lowest 1000 m AGL, the dotted line corresponds to the IWV above 1000 m AGL. 77

6.7 MSG SEVIRI 10.8 μm BT observations on 11 April 2006, 02 UTC. Dotted line corresponds to the approximate ITD position derived from Méso-NH results and ECMWF analyses by looking for the 15° dewpoint contour line. 80

6.8 Top: East-west averaged temperature at 10 m AGL (first Méso-NH level) as a function of latitude, averaged over six grid boxes between 1.55 °E and 2.0 °E at different times on 10 and 11 April 2006. Bottom: Observed MSG infrared BTs of 10.8 μm channel (averaged over 1.5 °E and 2 °E). Vertical dashed line indicates the position of Nangatchori. 81

6.9 Solid line: ITD position as a function of time over the whole Méso-NH latitudinal cross section (average over 1.6 °E to 2.1 °E) on 10/11 April 2006. The position was diagnosed from Méso-NH output by looking for the latitudinal position of the strongest cooling in 10 m within one hour (solid line with diamonds). Data are plotted at xx30 UTC, assuming that the ITD displacement speed does not change much within one hour. Dotted line with crosses: Time of maximum MSG BT gradient over the latitudinal cross section between 8 °N and 11 °N (average over 1.6 °E and 2.1 °E). Horizontal dashed line indicates the position of Nangatchori. 82

6.10 Méso-NH meridional cross sections, averaged over six grid boxes between 1.55 °E and 2.0 °E at different times on 10 and 11 April 2006. Vertical dashed line indicates the position of Nangatchori. Top: Wind speed in 925 hPa level. Center: Pressure difference relative to Nangatchori pressure value. Bottom: Integrated water vapor. 83

6.11 Meridional cross section of modelled 11 g kg^{-1} water vapor mixing ratio contour line at different times on 10 and 11 April 2006. 84

6.12 Meridional cross section of modelled water vapor mixing ratio at 1.8 °E on 11 April, 00 UTC. White contours represent zero meridional wind speed (which corresponds to the ITD position at the ground). 86

6.13 Model pseudo-soundings of potential temperature, water vapor mixing ratio, as well as the zonal (qu) and meridional (qv) water vapor advection fluxes over Nangatchori. 87

6.14 Statistical overview of Nangatchori measurements in April 2006 89

7.1 10/11 April 2006, ITD arrival. Top: BTs along the water vapor line at 70° elevation angle. Bottom: Same for 5.4° elevation angle (pointing to north). 94

105

List of Tables

3.1　Instruments in Nangatchori in 2006, used in this study. 23
4.1　Frequencies (in GHz) of the microwave radiometers deployed at Nangatchori and Niamey. 27
5.1　Seasons in 2006 as defined by Lothon et al. (2008). 47
5.2　Sub-periods of the monsoon season in 2006 as defined by Bock et al. (2008). 47
5.3　Mean IWV and standard deviation for different parts of the monsoon season 2006, defined in Tab. 5.2. 49
5.4　Monthly percentage of cloudy times observed by ceilometer (below 7000 m AGL) and rainfall (in mm) from rain gauge observations in Djougou and Niamey. 49

6.1　Typical diurnal cycle over West Africa from the Gulf of Guinea to the Sahel during April 2006. 85

.1　Data availability and weather conditions during June 2006. Cloud properties: ci=cirrus clouds, gf=ground fog, dc=daytime cumuli, al=aerosol layers, wcl=water cloud layers. Microwave properties: iwv=integrated water vapor, lwp=liquid water path, BT=bright band Codes: 0 - no data available, 1 - less than 50 % availability, 2 - some data gaps, but more than 50 % availability, 3 - 100 % availability. 110
.2　Same as Tab. .1, but for July 2006. 111
.3　Same as Tab. .1, but for August 2006. 112
.4　Same as Tab. .1, but for September 2006. 113

Acronyms

AEJ	African Easterly Jet
AGL	above ground level
AMF	ARM mobile facility
AMMA	African Monsoon Multidisciplinary Analyses
ARM	Atmospheric Radiation Measurement
BT	brightness temperature
CBL	convective boundary layer
CNRM	Centre national de recherches météorologiques
CNRS	Centre national de la recherche scientifique
DSD	drop size distribution
ECMWF	European Centre for Medium-Range Weather Forecasts
EOP	Enhanced Observing Period
FM-CW	Frequency modulated continuous wave
GPS	Global Positioning System
HATPRO	Humidity And Temperature PROfiler
IRD	Institut de recherche pour le développement
ITCZ	Intertropical Convergence Zone
ITD	Intertropical Discontinuity
IWV	integrated water vapor
LOP	Long-term Observing Period
LWC	liquid water content

LWP	liquid water path
MCS	Mesoscale Convective System
MRR	Micro Rain Radar
MSG	Meteosat Second Generation
MSL	mean sea level
NOAA	National Oceanic and Atmospheric Administration
NWP	numerical weather prediction
PBL	Planetary Boundary Layer
RMS	root mean square
RT	radiative transfer
SEVIRI	Spinning Enhanced Visible and Infrared Imager
SOP	Special Observing Period
SPI	Standardized Precipitation Index
UHF	Ultra High Frequency
UTC	Universal Time Coordinated
VHF	Very High Frequency
WP	Work package

Appendix

Tables on data availability

Table .1: Data availability and weather conditions during June 2006.
Cloud properties: ci=cirrus clouds, gf=ground fog, dc=daytime cumuli, al=aerosol layers, wcl=water cloud layers. Microwave properties: iwv=integrated water vapor, lwp=liquid water path, BT=bright band
Codes: 0 - no data available, 1 - less than 50 % availability, 2 - some data gaps, but more than 50 % availability, 3 - 100 % availability.

Instrument		Lidar ceilometer Vaisala CT25K		Microwave Radiometer HATPRO		Micro Rain Radar MRR		Precipitation Raingauge
Date	Code	Comments	Code	Comments	Code	Comments		in mm
1. Jun. 06	3	fog, ci	0		3	22-24 low		0.0
2. Jun. 06	3	ci, dc	2	short rain	3	16 short		0.0
3. Jun. 06	3	few ci and dc	3	short rain	3	3.5 short		0.7
4. Jun. 06	3	al 2000 m, dc	3	dc, IWV rise	3	no rain		0.0
5. Jun. 06	3	gf, strong PBL rise	3	air mass change 6am	3	no rain		0.0
6. Jun. 06	3	gf, dc	3	very few dc	3	no rain		0.0
7. Jun. 06	3	al 3000 m, few dc	3	very few dc	3	no rain		0.0
8. Jun. 06	3	ci, PBL rise	3		3	no rain		0.0
9. Jun. 06	3	wcl > 4000 m, rain, ci, gf	2	afternoon rain	3	12.5-18.5 BT@4km		12.3
10. Jun. 06	3	ci, al up to 300 m	2	calibration	2	no rain		0
11. Jun. 06	3	al,gf, PBL rise, rain,ci	3	IWV rise,	3	21		0
12. Jun. 06	3	PBL rise	2		3	no rain		0
13. Jun. 06	3	al lowering, PBL rise !!	3	no clouds	3	no rain		0
14. Jun. 06	2	rain, ci	2	rain in early morning	2	2-6,5		7.9
15. Jun. 06	3	many clouds > 3000 m	3	dc with LWP, showers	3	9, 12		0.6
16. Jun. 06	3	evening wcl 1500m	3	clouds in night	3	23		0
17. Jun. 06	3	al 2000 m, rain	2	early morning rain	2	5,5 - 7,5		17.3
18. Jun. 06	3	wcl, rain, later wcl 6000m	3	daytime rain events	3	7, 10-13		0.5
19. Jun. 06	3	PBL rise	3	dc	3	no rain		0
20. Jun. 06	3	PBL rise, high wcl/ci	3	many clouds	3	no rain		0
21. Jun. 06	3	gf, dc	3	dc	3	no rain		0
22. Jun. 06	3	rain, ice clouds	3	early morning rain	3	3 - 5		3.5
23. Jun. 06	3	gf, rain, some wcl	3	early mor.; noon rain	3	4 - 5 ; 12- 15		3.2
24. Jun. 06	2	dc	2	2 strong clouds	3	no rain		0
25. Jun. 06	3	few wcl	3	one big cloud	3	no rain		0
26. Jun. 06	3	al 1500m !!, dc	2	few clouds	3	no rain		0
27. Jun. 06	3	al, gf, PBL rise, dc	3	few clouds	3	no rain		0
28. Jun. 06	3	al, PBL rise	3	dc	3	no rain		0
29. Jun. 06	2	rain, dc, wcl	2	rain between 1-2, dc	2	1 - 2		23.8
30. Jun. 06	2	gf, rain, ci	3	morning rain	3	8 - 9-5		3.3

Table .2: Same as Tab. .1, but for July 2006.

Instrument	Lidar ceilometer Vaisala CT25K		Microwave Radiometer HATPRO		Micro Rain Radar MRR		Precipitation Raingauge
Date	Code	Comments	Code	Comments	Code	Comments	in mm
1. Jul. 06	3	dc, PBL rise, wcl 5000	3	clouds	3	17	0
2. Jul. 06	3	dc, wcl	3	clouds	3		0
3. Jul. 06	3	rain, PBL rise, wcl	3		3	1,5-2,5 ; 14-15 ; 17	12.9
4. Jul. 06	3	al, dc, rain shower	3	afternoon clouds	3	16.5 ; 21	2
5. Jul. 06	3	dc	3	afternoon clouds	3	no rain	0
6. Jul. 06	3	gf, PBL rise, some wcl	3	clouds	3	0 shower	0
7. Jul. 06	3	dc	3	morning clouds	3	no rain	0
8. Jul. 06	3	gf , sc ci, al	3	daytime clouds	3	no rain	0
9. Jul. 06	2	al, sc, rain?	2	1 bifg cloud	2	13	0
10. Jul. 06	3	wcl 4000, PBL rise	3	afternoon clouds	3	no rain	0
11. Jul. 06	3	gf, PBL rise	3	many clouds	3	no rain	0
12. Jul. 06	3	gf, PBL rise	3	cloud, IWV min 19	3	no rain	0
13. Jul. 06	3	gf, PBL rise	3	morning clouds	3	no rain	0
14. Jul. 06	3	al, gf, dc	3	afternoon clouds	3	no rain	0
15. Jul. 06	3	many cloud ; rain, all z	3	clouds ; rain	3	4-7 ; 13-16	21
16. Jul. 06	3	dc, high clouds	3	few clouds	3	no rain	0
17. Jul. 06	3	rain, ci	2	rain	3	8-13 ; 19	15.7
18. Jul. 06	3	wcl 5000m, PBL rise	3	evening clouds	3	no rain	0
19. Jul. 06	3	many clouds ; rain	2	rain	3	0-2 ; 16,5-24	70.1
20. Jul. 06	3	sc, rain	2	IWV decreases	3	low: 0 - 5 ; 11-14	1.8
21. Jul. 06	3	4000 m clouds, dc, rain	2	high LWP	3	20-24	44.7
22. Jul. 06	3	rain, gf, wcl	1	rain	3	0-2; 10- 12 ; 14-17	12
23. Jul. 06	3	2 level clouds	2	good data > 10	3	6 low	0
24. Jul. 06	3	gf, rain, ci	2	afternoon rain	3	13,5 17-19	0
25. Jul. 06	3	gf, PBL rise	2	wet	2	11,5; 15; 17 -18	7
26. Jul. 06	3	fog, PBL rise !! 2?	3	IWV decreases	3	no rain	0
27. Jul. 06	3	wcl 5000m sc	3		3	strong shower 23	1
28. Jul. 06	1	gf-rain	1	power cut after 6	1	strong shower 6	34.4
29. Jul. 06	0	power cut	1	after 15	1	after 15 no rain	0
30. Jul. 06	1	sc, rain, wcl 4000 m	2	rain	3	18-21	7.6
31. Jul. 06	3	wcl 4500 m, sc	3	cloud at noon	3	shower 14	0

111

Table .3: Same as Tab. .1, but for August 2006.

Instrument	Lidar ceilometer Vaisala CT25K		Microwave Radiometer HATPRO		Micro Rain Radar MRR		Precipitation Raingauge
Date	Code	Comments	Code	Comments	Code	Comments	in mm
1. Aug. 06	3	gf, PBL rise, rain	3	daytime clouds, rain	3	20-22 ; 24	23.8
2. Aug. 06	2	rain, wcl 4000, sc	2	rain, some clouds	2	0-2	4.2
3. Aug. 06	3	rain, sc, ci	3	morning dayclouds, rain	3	2 (light)	0
4. Aug. 06	2	gf, sc, rain, many wcl	2	daytime clouds, rain	3	21-24 (light)	0
5. Aug. 06	3	rain, gf, PBL rise, ci	3	clouds, rain	3	2-4, 18-21	3.7
6. Aug. 06	1	gf	3	morning rain, many cl	3	10-12	0
7. Aug. 06	3	gf, dc, PBL rise	3	afternoon clouds	3	no rain	0
8. Aug. 06	3	gf, PBL rise, sc	3	afternoon clouds	3	no rain	0
9. Aug. 06	2	gf, PBL rise	2	afternoon clouds	3	no rain	0
10. Aug. 06	2	gf, wcl 2000, rain	2	IWV rise	2	16-17 (shower)	10.2
11. Aug. 06	2	PBL rise, rain, wcl 5000	2	rain, clouds	2	15-19	17.8
12. Aug. 06	2	gf, PBL rise, rain	2	many clouds, rain	2	16-20	25.9
13. Aug. 06	3	gf, PBL rise, rain	3	daytime clouds, rain	3	17-19	8
14. Aug. 06	3	gf, PBL rise	2	daytime clouds, rain sh.	2	16 (shower)	0
15. Aug. 06	2	gf, dc, rain	2	daytime clouds, rain	2	13-14 (light)	0
16. Aug. 06	3	many sc, rain	2	clouds, evening rain	3	16 (shower)	9.9
17. Aug. 06	3	gf, PBL rise, ci, rain	3	daytime clouds, rain	3	14-15, 17-20	4.2
18. Aug. 06	2	gf, wcl 2000	2	many clouds	3	no rain	0
19. Aug. 06	3	gf, sc, PBL rise	2	morning clouds ; rain	2	4	0
20. Aug. 06	3	gf, PBL rise	3	IWV decrease, some cl.	3	no rain	0
21. Aug. 06	3	gf, PBL rise	3	afternoon rain	3	17	0
22. Aug. 06	3	gf. PBL rise, rain, ci	2	IWV rise, rain	3	17-20	6.4
23. Aug. 06	1	gf, rain	2	much rain	3	5-7 ; 13-17 ; 19-23	83.9
24. Aug. 06	1	wcl, ci	2	many clouds	2	no rain	0
25. Aug. 06	3	gf, PBL rise, rain	3	afternoon clouds, rain	3	16-17	0
26. Aug. 06	2	dc, wcl 4-6 km, rain	2	daytime clouds, rain	3	15-17	22.7
27. Aug. 06	0		3	early morn. clouds, rain	3	2, 7-8	0
28. Aug. 06	0		2	rain, morn. Clouds	2	10-12	3.9
29. Aug. 06	2	dc, sc !	2	many clouds	2	10 (shower)	0
30. Aug. 06	2	PBL rise, wcl, rain	2	rain,	2	6-7 ; 12 ; 15-18	2.7
31. Aug. 06	1		1	rain	1	15, 18-23	7.6

Table .4: Same as Tab. .1, but for September 2006.

Instrument		Lidar ceilometer Vaisala CT25K		Microwave Radiometer HATPRO		Micro Rain Radar MRR		Precipitation Raingauge
Date	Code	Comments	Code	Comments	Code	Comments		in mm
1. Sep. 06	1	wcl 6000, sc	2	rain, clouds, wet radome	2	8 (light)		0
2. Sep. 06	3	night stratus, dc	3	daytime clouds	3	no rain		0
3. Sep. 06	3	rain, wcl 5000	3	rain, clouds, wet radome	3	9-13		27.2
4. Sep. 06	2	many clouds, gf	2	many clouds, IWV rise	2	14 (shower)		0
5. Sep. 06	3	gf, PBL rise, wcl, rain, ci	2	rain, daytime clouds	3	14-18		3.9
6. Sep. 06	3	dc, gf, rain	2	afternoon clouds, rain	3	20-22		3.1
7. Sep. 06	3	dc, rain, ice	2	rain, daytime clouds	3	15-18		2.9
8. Sep. 06	3	gf, PBL rise, rain, al	2	daytime clouds, rain	3	16-18		1.3
9. Sep. 06	3	gf, PBL rise, rain	2	rain, many clouds	3	0-1, 15, 17-20		24.1
10. Sep. 06	2	gf, dc	2	many clouds	2	17 (light)		0
11. Sep. 06	2	dc	2	daytime clouds	2	no rain		0
12. Sep. 06	2	rain, gf, PBL rise	2	much rain	2	1-10, 17-22		54.2
13. Sep. 06	0		2	rain	3	13 (shower)		0
14. Sep. 06	2	sc	0		3	no rain		0
15. Sep. 06	3	rain, ci, gf	0		3	11-15		8.9
16. Sep. 06	3	wcl 4000, gf, dc	0		3	no rain		0
17. Sep. 06	3	sc, dc, rain	0		3	3-5		0.5
18. Sep. 06	3	sc, dc	0		3	no rain		0
19. Sep. 06	2	wcl 5000, PBL rise	1	rain	2	2-4, 17-19, 21		23.2
20. Sep. 06	1	sc	2	afternoon clouds	2	12 (shower)		0
21. Sep. 06	3	wcl 4000, gf	3	morning clouds	3	7-8 (showers)		0
22. Sep. 06	3	gf, PBL rise, rain	2	afternoon clouds, rain	3	13, 17, 20-24		38.2
23. Sep. 06	3	gf, PBL rise, dc	2	rain, afternoon clouds	3	2-7		1.1
24. Sep. 06	3	sc	3	morning clouds	3	no rain		0
25. Sep. 06	3	dc, rain, ci	2	evening rain	3	16-20, 23		35.9
26. Sep. 06	3	gf, rain, dc	2	daytime clouds	3	2		1.5
27. Sep. 06	3	rain, gf, PBL rise	2	rain, daytime clouds	3	0-1, 4-5, 16-18		9.4
28. Sep. 06	3	gf, sc, dc, rain	2	rain, daytime clouds	3	13 (light)		0
29. Sep. 06	3	gf, PBL rise	3	clouds, IWV rise	3	no rain		0
30. Sep. 06	1	wcl 4000, gf	1		1	no rain ?		3.9

Acknowledgments

This work was funded by AMMA, a project in the European Community's Sixth Framework Research Program. Based on a French initiative, AMMA was built by an international scientific group and is currently funded by a large number of agencies, especially from France, UK, USA and Africa. It has been the beneficiary of a major financial contribution from the European Community's Sixth Framework Research Program. Detailed information on scientific coordination and funding is available on the AMMA International web site http://www.amma-international.org.

I would like to express my gratitude to all people who contributed to the success of this work. First of all, I would like to mention Prof. Dr. Susanne Crewell who offered this PhD post in the frame of the AMMA project in early 2005. During the years, she always had time for discussions, and she gave me a lot of advice for my work.
I would like to thank all the other members of our working group "Integrated Remote Sensing" for their help at many different questions that arose, e.g., on programming, or on microwave radiometry, and for proof-reading of this manuscript.
Special thanks go to Dr. Thomas Rose from Radiometer Physics GmbH. He explained to me the measurement principles of the HATPRO radiometer and was always helpful when I had problems with the instrument and its software.

I would like to thank Diana Bou Karam from Service d'Aéronomie (CNRS) in Paris for providing the Méso-NH model results. I appreciated very much this collaboration. Many thanks go also to Dr. Cyrille Flamant, Dr. Olivier Bock and Dr. Frédérique Saïd for their comments and fruitful discussions on this work.

The first part of the work for this thesis consisted of installing and operating instruments in Benin which gave me the unique chance of travelling there five times and getting to know the country and its inhabitants. Despite many adverse infrastructural conditions, the observations were very successful which was only possible with the help of many people in Benin. First, I would like to mention Erik Houngninou who regularly checked our instruments and stored the measured data to DVDs which were then sent by snail mail to Europe—and all data arrived safely and unharmed! I want to thank Armand Mariscal who was the responsible person for the Nangatchori site and helped together with his colleagues from IRD in many logistical issues, such as transport of instruments as well as organizing very different things we needed for the installation like tubes, wires, desks and many other things. It was also very nice that I could stay overnight at the

IRD guest house in Djougou during my field trips. I'm very grateful to the people in the AMMA, IRD, and Impetus offices in Cotonou who solved many problems that arose before and during my stays in Benin, concerning transportation, customs, and many more. A special thank goes to the local guard of the Nangatchori site who permanently took care that the instruments were not harmed by cattle herds.

And finally, a big thank you to my family and especially my parents who always supported me and encouraged me to continue my way.

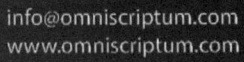

Printed by Books on Demand GmbH, Norderstedt / Germany